Geographic Information Systems Demystified

For a listing of recent titles in the *Artech House Mobile Communications Series*,
turn to the back of this book.

Geographic Information Systems Demystified

Stephen R. Galati

ARTECH
HOUSE

BOSTON | LONDON
artechhouse.com

Library of Congress Cataloging-in-Publication Data
A catalog record for this book is available from the U.S. Library of Congress.

British Library Cataloguing in Publication Data
A catalogue record for this book is available from the British Library.

ISBN-10: 1-58053-533-x
ISBN-13: 978-1-58053-533-5

Cover design by Robert Pike

685 Canton Street
Norwood, MA 02062

10 9 8 7 6 5 4 3 2 1

To my wife Janet and my children, Zachary, Nicholas, Sarah, and Jacob

Contents

Preface

The field of geographic information systems (GISs) is emerging as one of today's most exciting and progressive technical areas. GIS technology has evolved during the past four decades and is only now starting to penetrate major industry and service hubs. With such a wide array of new users, there exists not only considerable interest in GIS but also a fair amount of uncertainty and misunderstanding surrounding the discipline.

Like other technical fields, there have been various methods employed to convey the fundamentals, implementation, and importance within the industry. My personal approach within this book is to encourage the highest quality understanding of GIS fundamentals before suggesting the reader study through application. In taking this route, this book attempts to disband user uncertainty and misgivings through guiding discussions about GIS rudiments and technical theory. With this in mind, I want to clarify the book's distinct structure, level of technical information, and somewhat hands-off approach.

The book is structured into three distinct parts:

- Part I: What Is a Geographic Information System?
- Part II: Geodesy, Earth Models, and Coordinate Systems
- Part III: GIS Applications and Environments

Although the specific content of each division is summarized in the Introduction, I want to offer some insight into each part's overall intent and degree of technical detail. On the surface, the technical level varies among the parts, but with very good reason. Let me briefly explain.

Part I presents the preliminary background of what a GIS is, why it exists, how it works, and what forms of data it uses. Within this first division, my primary focus is to get readers up to speed on GIS's composition and I take an explicitly less technical path to achieving this. A more technical approach to this background information would leave the reader swathed in tangential theory and would needlessly add further complexity to an already intricate subject. The topics in Part I are supportive in nature and are approached as such with distinct technical brevity.

Part II is the true "nuts and bolts" of GIS and, as such, is approached with a much greater technical treatment. My approach to Part II is to comprehensively explain the theory behind GIS's core components, such as the reference model of the Earth, the coordinate systems, and the map projection. As you will notice, Part II is dramatically more complex in nature than Part I and this is done for a specific reason: Most GIS uncertainty, misgivings, and misuse extend from a limited understanding of these core components. I believe a comprehensive understanding of these topics is essential for effective, proper, and creative GIS use.

Part III takes a step beyond the theoretical to provide an overview of geographic data presentation, real-world applications, and inexpensive (practice) systems and data. To this end, Part III is technically less complex than the previous part for the sole reason of conveying a clearer vision of how GISs are actually used, rather than overwhelming the reader with a thorough discussion on each element.

I want to highlight one other feature—a benefit—to my approach: It is theory based rather than hands-on based. I call it a benefit because true hands-on approaches lead the reader to focus primarily on the use of a GIS (typically on a specific system) rather than on understanding a GIS. Presented too early, this hands-on method, I feel, becomes an impediment to comprehension. A quality theoretical understanding is the key ingredient to effective and insightful real-world use. A hands-on approach restrictively focuses on a particular system, while a theory-based approach lets the user (reader) freely experiment and choose the system on which to learn. This is not to say a hands-on approach to GIS is a wrong learning method; I am simply stating that for the best use of a GIS and to gain the greatest amount of user perspective, the theory-based approach proves beneficial and leaves the reader at the doorstep of hands-on practice.

I hope you enjoy this book and gain a deep appreciation and excitement for geographic information systems.

Acknowledgments

Many people provided their support, guidance, and encouragement during the research and writing of this book. I wish to express my thanks to the following:

Dr. Tyler S. Pruett, for his significant input in the book's overall structure and his help with the writing and technical review of some important chapters, particularly Chapters 1, 3, 9, and 12. I am grateful for his early guidance and support of this book.

Dr. Paul Aplin at the University of Nottingham, United Kingdom, for his comprehensive technical review and keen eye for my many errors. His superior technical and structural suggestions have enabled me undoubtedly to write a much better book.

René Essiembre, for his assistance and support during the writing of some seriously technical chapters. I am grateful he let me "pick" his GIS-brain.

I have also benefited from the help and support of John R. Kubasek from John R. Kubasek & Associates, Inc.; John Pavek and Glenn Small, PE from T&M Associates, Inc.; and Van Bowersox from the National Atmospheric Deposition Program (NADP).

I wish to express my tremendous gratitude to Mark Walsh and Barbara Lovenvirth from Artech House. Mark's support of this project has remained steadfast from the very beginning and throughout the thick and thin of the book's development. Barbara has been equally committed to this project throughout its development and has shown me remarkable support and guidance. I thank them wholeheartedly for believing in this project.

My heartfelt thanks go to my beloved family—my wife Janet, my youngest son Jacob, and my adoring triplets, Zachary, Nicholas, and Sarah. My family has inspired me to continue writing "the book" and has shown me much needed enthusiasm when mine was waning. Their support and love is endless as is my love for them.

Finally, and most importantly, I want to thank God for giving me the fortitude to complete this book.

Introduction

The 2004 U.S. presidential elections marked a milestone in how information is used and portrayed. For the first time ever, analysts used geographic information systems (GISs) to generate election maps that showed which presidential candidate was leading in each county in every state throughout the country. This information was well received by political analysts and home viewers alike, though few understood the *thing* called GIS. To most, the highly visual election results were the products of some "black box" that offered amazing analysis tools. Understanding how the "black box" worked was not their concern.

The perception of GIS in the business world goes much the same way. Many engineers, environmentalists, and business professionals who use GIS do not necessarily know how it works. This is not unfounded since groundbreaking technology typically evokes initial skepticism and uneasiness. Do you remember the initial unease associated with the "smart machine" in the early 1980s? Back then, personal computers were new and quickly embedding into the worldwide society. This "smart machine" apprehension resulted in the typical person using computers only for word processing and merely tapping a small percentage of the computer's true capability.

GIS is not secluded from these issues of technology fear. Actually, modern GIS environments are almost 100% dependent on the "black box." Without software technology, GIS as we know it would cease. Consequently, the GIS community experiences the same uneasiness among its users. Many GIS novices, and even some intermediates, use but a fraction of the functional capability of GIS software systems, minimizing the system's ability to that of limited queries and static mapmaking. Others politely refrain from using GIS at all, believing it will be complicated, overly technical, or a waste of time altogether.

The rapid growth of GISs worldwide, particularly in the 1990s, has challenged users with the void of GIS standards and software market saturation. Company after company has injected GIS software products into the open market, transforming the flux of GIS capitalism into a sequence of software start-ups and reinventions. Sadly, this has left users, both novice and advanced, apprehensive about what is the right software environment for their time and money investments. Only recently has this lack of standards become an evolving topic in the GIS community, to which the ultimate resolution will be a homogeneous GIS user interface.

Pushing beyond the apprehension to learn GIS requires not only graphical consistency in the GIS platform and environment but also the recognition that GIS is an iterative discipline. *Learn by doing* is a caveat to any valid GIS approach, although to get the most out of a GIS, the user must not only use the system but also understand its potential. Even a rudimentary understanding, such as a basic grasp of fundamental concepts, can intrinsically help with the conveyance, analysis, and final usage of geographic information. With understanding, these modern-day pioneers have at their disposal a powerful analysis tool that will open new avenues of data analysis in their applications. The success of GIS for the 2004 elections only proved that a basic understanding could provide amazing tools for analysis and real-time information conveyance.

GIS has been around for nearly four decades but has gained a true user backing only within the past 10 years. With technical and industry professionals becoming aware of GIS's capability and benefits, the dedicated following of developers and users is now growing. The next 10 years will mark new milestones in geographic information, while everyday life will continue to embed with GIS applications.

A Definition of Geographic Information Systems

So what is a GIS? A *geographic information system,* as defined in the Environmental Systems Research Institute's (ESRI's) *Dictionary of GIS Terminology,* is a collection of computer hardware, software, and geographic data for capturing, storing, updating, manipulating, analyzing, and displaying all forms of geographically referenced information. This is a very general description for such a complex and wide ranging set of tools. GIS is, in essence, a central repository of and analytical tool for geographic data collected from various sources. The developer can overlay the information from these various sources by means of themes and layers, perform comprehensive analysis of the data, and portray it graphically for the user.

In the case of the 2004 elections, a layer discerning counties overlaid a base map of the United States. County demographics were developed into the GIS prior to election day and integrated other geographic data sets such as candidate information. These data sets for county demographics are readily available.

Throughout election day, real-time data from each reporting precinct in every county fed into the GIS and was cross-referenced with the existing data. The result was three-dimensional mapping that illustrated not only candidate leaders and preelection poll accuracy in each county but also voter turnout by socioeconomic status and party affiliation.

GISs hold a wide range of functionality and analytical capability that with the right developer can provide a collection of useful data and maps. Although the method of presenting geographic information has transformed throughout the ages, the individual uses have only expanded. In fact, such yesteryear applications as farming, sailing, cartography, and warfare still use geographic information.

Poincaré's Maps and the Move Toward GIS

Jules-Henri Poincaré can be considered one of the first bridges between the old technology of cartography and the first traces of GISs. Poincaré was one of France's most innovative thinkers of the late nineteenth and early twentieth centuries. His work covered the landscapes of mathematics, time, physics, relativity, and geodesy. He used his geometric visualization methods, which portrayed non-Euclidean geometry through "visually" representative maps, to fill in the blanks on the world map.

Poincaré began his prestigious career examining coal mines throughout France. While working as a mining engineer, Poincaré was instrumental in determining the root cause for a deadly explosion of a mine pit in Magny. He conducted a meticulous and logical investigation of the events while maintaining consideration of the mine's structure, geology, worker locations, and reason. Poincaré drew highly detailed maps of the mine that indicated the flow direction of air through the mine shaft. This gathering of information on a single diagram assisted Poincaré in determining that an accidental puncture of a Davy "safety" lamp caused the ignition of the mine's resident methane gas. He also provided the exact location of the ignition and exact lamp punctured (lamp 476). After the Magny incident, Poincaré became reputed for his visualization techniques.

In 1889, Poincaré's notoriety spread with the presentation and acceptance of his visual mapping method. He plotted the habits of a small asteroid repetitively orbiting around Jupiter and the Sun (known as a three-body problem). He intentionally neglected the examination of the entire orbital path and, ingeniously, focused on the orbit's impact points within a single plane in space. The result was a stroboscopic diagram identifying a complex series of plotted impact points. This plotted diagram caused a stir in the scientific community because it identified chaos, or instability, within the natural process of the universe. The mapping method, now known as Poincaré's Map, intertwined geography, location, visualization, data, and analysis.

In the following years, Poincaré gained access to France's bureaucracy, acting in various high-level, high-visibility positions. As the Bureau of Longitude's scientific secretary, Poincaré became heavily involved with the escalating and competitive telegraphic longitude work. France had not made any important impacts as had other worldwide entities, such as the United States and Britain. France and Poincaré wanted to establish telegraphic simultaneity, a primitive methodology compared to today's global positioning system (GPS) technology. At the same time, the International Geodetic Association desperately needed an accurate length measurement of a meridial arc to remeasure the shape of the Earth. France jumped at the chance.

In 1900, Poincaré headed a mission to help France's Bureau of Longitude accurately plot the region of Quito, Ecuador. Although his expedition teams in Ecuador met misfortune every step of the way in the difficult, obstacle-filled terrain, Poincaré was successful. After 7 years, Poincaré completed the mission and created a network of latitude and longitude measurements that later were connected by telegraphic cable and linked to the world telegraph network. Poincaré, once again, used his visual geometrical methods to help detail the measurement data and cable pathways upon a geographic map.

Through his many exploits, Poincaré used mathematics and ingenuity to expand the parameters of map visualization, physics, geodesy, and scientific concept. In doing so, he revolutionized methods of using geodetic information to widen the realm of analysis. Undoubtedly, Poincaré left his mark as an early pioneer to modern-day information systems, such as GIS and GPS, and his Poincaré Map was a predecessor to present-day chaos theory.

Old Technology, New Platform

As seen through the work of Poincaré, the concept of a GIS is rather old. Ancient cartographers used observation, field informants, celestial positions, and measuring tools (such as the sextant) as inputs to calculate distances and determine land formations. Hand-drawn maps then depicted this information. Maps indicating demographics, such as the layout of crops throughout the nation or population, were based upon field information and generalization and were not completely accurate. Usually informed generalization and estimation helped develop the maps.

As technology increased, so did the precision of maps. Graphic landforms more accurately portrayed reality, while maps better portrayed demographics. Color palettes, symbol sets, and attribute tables depicted data.

Today, geographic information is at its most precise. As the new platform for real-time data, GIS utilizes high-resolution satellite imagery, light intensity detection and ranging (LIDAR), GPS data, computer-aided design (CAD) files, and enterprise database management systems for complex analysis. Users can cross-reference these geographic data sources in a GIS and create impressive geographic displays catered to an area of interest and to individual or specific applications. Users now have the opportunity to customize their maps not only to show the information they want but also how they want it. For instance, an engineering firm widening a section of interstate roadway not only can use a GIS to mark a satellite map of the roadway section with bridge locations, exit and on ramps, and elevations but can also map soil boring points with hyperlinked data, traffic data, surrounding vegetation types, and roadway drainage.

GIS is appearing in almost every industry, forging its place as the new platform for geographic information. GIS is now used for education, land management, natural resource management, environmental, and aeronautical applications, just to name a few. GIS even crosses every industrial and humanitarian threshold with software and data development efforts. The only limits to GIS are the user's fundamental knowledge of the system and overcoming the stigma of creative GIS use.

The GIS Mystique

Whether it is the result of users' anxiety, disinterest, or uncreative whim, GIS bears a cross of a subtle mystery. This GIS mystique, as I call it, is the ultimate result of misconception and limited knowledge. Let me explain.

GIS holds a preconception that is grossly different from reality. For instance, GIS is sometimes solely viewed as mapmaking software. This is simply not true. GIS has found its way into everyday life, from the internal management of utility companies that deliver electricity and water to mapping distant planets, such as the European and U.S. expeditions on and around Mars and Venus. The reality of GIS, as this book details, is that it can do so much more than make a map.

Similarly, GIS is sometimes wrongly linked on a one-to-one basis with GPS technology, found in handheld receivers, E911 systems, and cellular phones. This user misconception "dumbs down" GIS technology to the point that it severely limits the ultimate capability and potential of a properly structured system. GPS data is but a small portion of the information a GIS manages and, in all honesty, is not needed for every user's application.

To overcome this communicable mystique, we must first understand that there is still not enough hands-on work with GIS by all levels of users. The users are only interested in information at their fingertips from the "black box," while management rarely understands the levels of visualization in GIS or how it can be successfully implemented. Consequently, many GIS development efforts are underfunded and prematurely abandoned.

A properly structured GIS requires time, and many users are either not willing to invest this time and money or not afforded the opportunity to design a system that maximizes capability. As competition increases, more users hastily implement a GIS to stay in line with their market and, subsequently, promote the fact they use GIS technology. If truth were told, many industry people want the term "GIS" on their laurel list, though they have little intention of using the software. Many systems suffer from this lack of use and grow stagnant. Competition, along with users' indolence, dilutes GIS structure and scientific implementations.

About This Book

Geographic Information Systems Demystified is written to give readers a digestible approach to understanding what a GIS is and how it can be used, while revealing the falsehoods of the GIS mystique. This book is not all encompassing but rather focuses on GIS theory. It presents the important elements of GIS to get anyone started and ready to work with certainty, and provides candid insight into this often unfamiliar technology science.

While many GIS learning sources are software product biased, the book offers an unbiased approach to the fundamentals of learning GIS, helping readers become functional and knowledgeable users. This guidebook also presents an original, comprehensive, and innovative treatment of spatial coordinate systems, a topic rarely covered at all in most popular GIS learning titles. The book examines and explains critical GIS concepts in a consistent structure, demystifying an often misunderstood discipline.

The structured topics first lay down the basics of GIS and then build upon this knowledge foundation with more specific information. The book is laid out in three parts:

- Part I: What Is a Geographic Information System?
- Part II: Geodesy, Earth Models, and Coordinate Systems
- Part III: GIS Applications and Environments

Part I focuses upon what a GIS is, why is it needed, the structure of geographic data, and the importance of metadata. It offers an understanding of how geographic data transforms into geographic information through a succinct presentation of geographic information basics and solidifies the knowledge foundation for the entailing topics found in Part II.

In Part II, the discussions tackle the more complex elements of GIS applications and, quite different from Part I, are much more technical and comprehensive. Core topics, such as map scale, Earth models, and map projections, are scrupulously defined and add to the host of GIS skills designed to give the user creative control over the analysis, portrayal, and use of data. The most common GIS tools are discussed in technical detail geared to get anyone quickly up and running with their own GIS applications.

Part III highlights some of the creative tools and techniques presently available within the discipline of GIS, as well as how GIS is used today in various industries. It details how geographic data can be uniquely used for analysis and presentation. Topics such as thematic mapping and real-world examples offer the user ways to creatively depict their geographic data and widen the spectrum of possible GIS applications. Also addressed are places to get free GIS environments, component software, and geographic data.

This book includes useful appendices and a glossary that may be used in tandem with the text. This information offers handy standards for ellipsoid parameters and datum transformation tables to use. This information was useful in the writing of this book and I felt it would be equally useful to the readers. The glossary defines words and phrases used throughout the book but should not be viewed as a comprehensive glossary of GIS terminology. If you wish for a more thorough glossary, check out some of the resources in the bibliography.

The Book's Approach to GIS Concepts

With GIS's realistic business and engineering models offering attractive cost savings and efficiency, geospatial systems can sometimes transcend practical concerns. The technology injects a touch of creativity into the veins of global infrastructure and serves decision makers on all levels of business and financial permanence. Important new GIS concepts are born every day, and industry trends fluctuate so frequently that a tempered approach to introducing GIS concepts hardly seems fitting. This book attempts to show GIS science and industry as they are, providing a literal roadmap through the industry's landscape.

We must remember that GIS is still brand new to the overwhelming majority of desktop computing users, as well as the technical and business fields. The importance of understanding fundamental GIS concepts is clearly underestimated in the industry today. This book addresses the most important concepts in earnest and details methods for using and developing geospatial systems. Every critical concept is discussed in a clear, consistent voice that is intended to raze any false preconceptions the reader may have.

Since GIS is an international infrastructure organization tool, the book takes a worldwide approach to the technology. GIS is booming throughout the world, especially in Europe, the Americas, India, Japan, and Australia. Topics are designed to further the overall understanding of GIS as it relates to worldwide applications.

The Goal to Demystify GISs

This book's main goal is to take the fear of the "smart machine" out of the user and expose the inherent mystery of GIS. Geared to the beginner and intermediate user, this book intends to forge a paced, seemingly trolling, pathway to productive understanding and unveil the wide range of user choice and creativity.

GIS is being used in almost every industry and affects some portion of every person's daily life, usually without awareness. Whether you realize it or not, your life is affected on some level by GIS technology. GIS is currently being used in almost every conceivable industry. Major GIS software companies boast more than a million daily application users. In fact, if you have ever used MapQuest, you have explored geographic data and performed analysis using GISs.

Geographic Information Systems Demystified intends to herald the broad availability of worldwide geographic information, diminish apprehension of the GIS unknowns, and widen the user's eyes to the great possibilities that can be achieved using GIS. In essence, this book aims to rip open the top of the "black box" and reveal the wonderful things that lie hidden inside.

Part I:
What Is a Geographic Information System?

From Geographic Data to Geographic Information

1

How Does a GIS Work?

In a geographic information system (GIS), geographic data are transformed into geographic information. This simple transformation, however, involves a complex series of functions and processes. In a nutshell, geographic data begins as raw positional feature data holding attributes. These data are then overlaid with complementary and/or contrasting data sets, which form coincident relationships. Data and relationships are analyzed, geoprocessed, and then presented as geographic information products. These geographic information products are often interactive software applications used to help people make decisions.

GISs are accessible to an array of users, from the expert GIS software developer to the GIS novice project manager, and, subsequently, offer visualization to users throughout the spectrum of skill levels. This diversity becomes a unique benefit of GIS and explains how quickly it can become visible within an organization, as well as offering project visibility to the public.

To fully understand the power of a GIS, we must first take a closer look at how raw geographic data becomes usable geographic information. The following sections depict the movement and manipulation of geographic data sets, as well as the areas where user ability is vital to usable output. The core geographic data transformation involves the following fundamental flow of information.

The Fundamental Flow of Information

Geographic data originate from actual locations and physical characteristics of features on or near the surface of the Earth (or other celestial bodies such as Mars or the Moon). These raw, positional data are the start points for every GIS by providing the basic geographic information needed for attribution, dataset modeling, relationships, and analysis. Refer to Figure 1.1. Raw data can come from a range of sources, such as aerial photographs, previously digitized maps, and global positioning systems. These types of digital and nondigital geographic resources are readily available and, in many cases, are plentiful, well designed, and comprehensive. Additionally, many digital sources are free. Other more labor intensive sources are field data and measurements collected from site visits, and transformed maps, whereby old hardcopy maps are first scanned into a computer and then digitized.

In the most fundamental sense, raw data are either geographic data or are transformed into geographic data through a GIS, and are in turn used to produce geographic information products. Geographic information products are user-conceived information results created through a GIS and a user's ability to relate, manipulate, and present overlaid geographic data. These products are used to analyze data for a specific application.

The overall work of transforming data in a GIS can be summarized through its three distinct procedures:

- *A GIS leverages the flexibility of geographic data.* Raw data are static (nonchanging) and offer only a limited amount of flexibility on real-world applications. When raw data are transformed to geographic data through a GIS, the capability for enhanced data use and analysis (i.e., data flexibility) significantly increases. At a minimum, overlaying two geographic data sources provides sufficient new information, better analytical means, and additional flexibility to not only help someone visualize the real world but also help them make an informed decision.

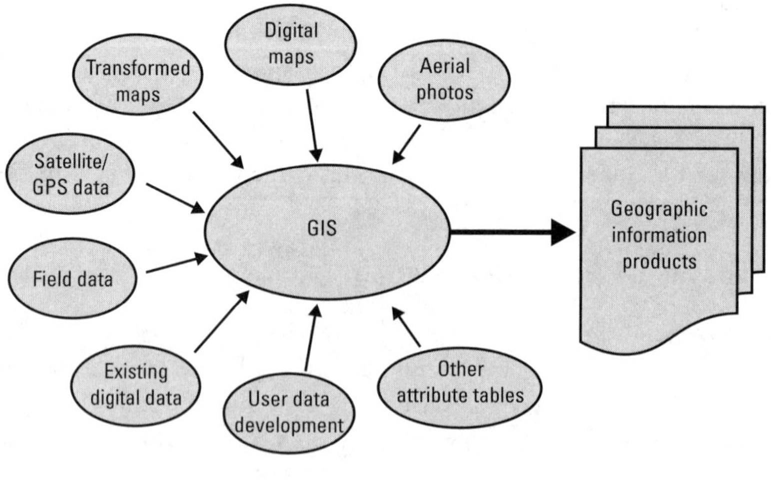

Figure 1.1 The path of geographic information.

- *A GIS performs functions and analysis within a single environment.* These data functions are the literal "doers" of a GIS solution and are known as *geoprocessing* and *spatial analysis.* These operations are available within a single GIS environment and include the generation of features, buffers, view sheds, and cross sections; the calculation of centroids, slopes, statistics, and suitability; and the manipulation of feature attributes, smoothing of lines, feature transformations, and clipping. Both geoprocessing and spatial analysis are discussed in more detail later in this chapter.
- *A GIS serves as a software application and creates useful information products.* GIS environments, foremost, serve as spatially enabled enterprise data management systems and data repositories. These systems are software applications that protect the value and usefulness of the information related to your project. The end result is an information product that enables the user to better manage his or her project.

Through this procedural flow of information, geographic data are transformed into geographic information. GIS environments centralize both data collection and information management to save time, minimize technical effort, and automate known repetitive administrative tasks.

The core data component of a GIS is often represented by a *geographic data model,* which is an industry or discipline-specific template for geographic data. A geographic data model offers the user flexibility in the design of the file management and database hierarchy. Geographic data models typically utilize a grid-based structure (known as raster) or a coordinate point structure (known as vector). Both models are explained in detail in the upcoming chapters.

Facilitating the model is one or more geodatabases. A *geodatabase* is a collection of geographic data sets, real-world object definitions, and relationships. Comparable to a Microsoft Access file, a geodatabase is a collection of geographic data sets and geometric features. A geodatabase furnishes the data organizational structure and workflow process model for the creation and maintenance of the core data product. In essence, the geodatabase is the heart of a GIS's management capability.

Let us now look at each element in the flow of information. We will start at the beginning with the various forms of raw geographic data and work our way to the GIS products.

Geography and Geographic Data

Geography is the study of the Earth's surface and climate, and is the founding science to a GIS. Geography furnishes information about the Earth and distinguishes how features upon the Earth correlate with one another. For example, a basic geographic study involves how climate and landform interrelate with inhabitants, soil, and vegetation. Data collected from this study are geographically oriented and are therefore geographic data. Any study with a geographic component, regardless of form, produces geographic data.

By their very nature, geographic data comprise the physical locations of objects on or near the surface of the Earth. Data are intimately concerned with the properties of such objects and hold attributes that can be associated to other types of geographic data. For example, a user can have two types of geographic data about a 40-acre stretch of land in New Zealand: one detailing elevation above sea level, and one detailing the various types of soil composition throughout the parcel. Both forms of data can be combined for analysis in a GIS using the land parcel as the common link.

All physical relationships between layers of geographic data are interpreted as coincident relationships, meaning that features coincide in real world space. In the above example, the elevation data and soil composition data would go on separate "layers" in the GIS. Layers are discussed later in this book.

Many times, geographic data are modeled in what is known as vector space. A vector space is simply a platform for geographic vector data, which use *x-y* coordinates with lines and shapes to depict Earth features. Geographic vector data store nontopological coordinate geometry and attribute information for spatial features.

Most standard GIS vector file formats consist of a feature file, an index file, and a linked attribute table. A *feature file* contains geographic object feature information, such as representative point, line, and polygon information. An *index file* contains unique identifiers that comprise more detailed information and help speed spatial feature queries. A linked *attribute table* is a matrixed table that contains explicative attributes for a group of spatial features.

The index file links the feature file data to the attribute table. Consequently, attributes and features exist in a strict one-to-many relationship, whereby geographic features can have multiple attributes. The vector model is readily recognized by those who use computer-aided design (CAD) environments, using feature geometry in a GIS vector model to instantiate points, lines, and polygons as objects. Figure 1.2 details sample vector data.

In most cases geographic vector data are *discrete data* occurring in cases where there are well-defined boundaries for physical representations or limited data values. The above figure involves tax lots that have visibly evident and discrete boundaries. Vector data can be created or modified directly by digitizing features using GIS development environments.

There is another major type of geographic data known as raster data. Raster data are digital images represented by a grid of valued pixels, or *cells*. The image type and the number of colors represented determine the properties and

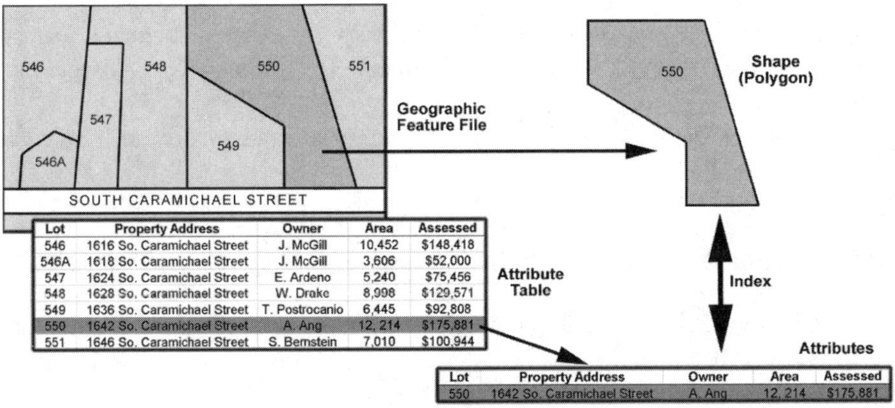

Figure 1.2 Vector data.

appearances of these pixels. Figure 1.3 details sample raster data that utilizes three different cell colors for three different property types.

In most cases, raster data are more suited for representing *continuous data* than vector data. Continuous data are a numeric form of data usually associated with the physical measurement of boundaries that are not well defined. Additionally, the surfaces represented are, in many cases, estimated through statistical

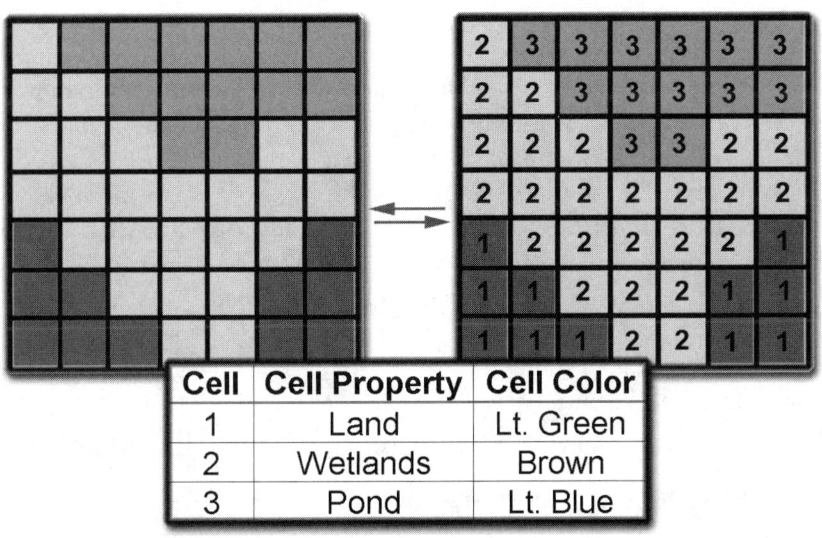

Figure 1.3 Raster data.

analysis. Raster imagery is perfectly suited for GIS sensor webs, a current industry trend that harnesses real-time observation and measurements over large, even global, regions.

Chapter 3 focuses specifically on the raster and vector data and offers a comprehensive discussion of each data model.

Georeferencing

Many data sources lack formal spatial referencing. Some CAD and GIS data sets are developed in a generic "design" space and have unique, often proprietary, types of referencing that simply need reinterpretation to be spatially integrated into a GIS environment. However, many of these sources are scanned raster data (digital imagery) that have only the coordinates of a raw pixel grid from an original scan. While these raster sources are often times unique and critical to a GIS project, images also need to be referenced from scratch, spatially transformed into a defined coordinate referencing system, then integrated and overlaid in a GIS environment. This process is known as *georeferencing.*

The ability to perform accurate and timely spatial referencing adds a measure of customization to any GIS operation or project. Raster imagery, such as hardcopy maps and aerial photography, is the most popular type of data to use with georeferencing, since it is the most commonly available type of data to use. Scanning imagery also alleviates the need to perform time-consuming and repetitive digitization efforts (i.e., transforming hardcopy to an electronic, digital file).

Georeferencing is the art of selecting common point locations in the real world using at least two data sources: an unreferenced source (such as a raster map) and a referenced source of the same area providing positional information. Basic georeferencing procedures involve point selection and transformation. For example, when a hardcopy map is scanned to an electronic file, it has no relationship to any real-world coordinate system. The georeferencing process establishes (or in some cases reestablishes) the relationship between image pixel locations and real-world locations. Georeferencing is accomplished by first selecting points on a source image (scanned raster map) with known coordinates for the real-world surface location (benchmarks, grid ticks, road intersections, and so on). These real-world coordinates are then linked to the corresponding pixel grid coordinates in the raster source image. After the image is georeferenced, each pixel has a real-world coordinate value assigned to it.

Queries: Locations and Attributes

Once data are adequately georeferenced and resident in a GIS, the user can then create a query expression to find the relevant data for a specific application. A query enables the user to search geographic data to collect location, feature, and attribute information from a relational database management system or geodatabase. The most fundamental test of geographic coincidence between data sources is a query by location in tandem with a targeting expression, such as *Intersection, Within distance of, Contained by, Share features with, Touch the boundary of, Are crossed by the outline of,* and *Completely in.*

Queries can also be executed on feature attribute tables and performed on specific fields. Records are selected in this manner; each record (or "row") in the table corresponds and is linked to a distinct geographic feature data set. Features conforming to certain attribute criteria can be selected and extracted using this method.

Quite simply, queries are the user's refining tool for taking the massive available data and selecting only those pieces that pertain to the application at hand. The absolute function of queries is a GIS staple and, subsequently, will be highlighted both in example and in operation throughout this book.

Geoprocessing

Geoprocessing is the fundamental process of creating a derived set of geographic data from various existing data sets using operations such as feature overlay and data conversion. In a typical geoprocessing environment, the user applies GIS functions to a group of geographic (input) data to yield a precise output data set suitable for a particular application. Geoprocessing functions run the gamut from simple spatial clipping to more complex analytic operations. These software functions can stand alone or be chained to other processes. This ultimately opens the gates for virtually unlimited sets of geoprocessing models and potentially staggering sets of output data to solve specific problems.

Most professional GIS software environments include a mission-specific geoprocessing interface, or "workbench," of geospatial dialogs and tools. These software environments usually include extensible scripting tools and compilers to automate, customize, and document geoprocessing workflows. The most important contribution of geoprocessing to the GIS big picture is the automation of repetitive tasks. Geoprocessing is an elaborate turnkey for efficient and clean geographic output.

Geoprocessors come in different forms. Many geoprocessing functions are embedded in a GIS environment. A GIS environment is a package of integrated GIS components: a geographic map control, a map layout designer, a data tree catalog, and so forth, of which geoprocessing is a member. However, many powerful stand-alone software applications offer specific, related subsets of geoprocessing functionality. Some include file format translators or spatial referencing transformation tools. Many professional GIS efforts actually require licensing these stand-alone accessories in order to reap the often-advantageous outputs of these stand-alone software applications.

Site selection is a prime example of a geoprocessing application. Another is the function of batch processing in a noninteractive manner. Geoprocessing environments can be considered "robots" that automate geographic data processes and provide storage of geographic data models. Geoprocessing is very reliable. In fact, within the GIS community geoprocessing environments are viewed as software-based insurance policies for the analyst trying to deliver geographic information products on time and within budget.

There are eight categories of predominant geoprocessing operations, or families of operations:

1. *Conversion.* Conversion is completely an issue of formatting. File format conversions (translations) and coordinate system referencing conversions are the most common geoprocessing conversion operations, and serve to characterize the conversion family.
2. *Overlay (union, intersecting).* Overlay involves superimposing two or more geographic data layers to discover relationships. In fact, overlay is intimately associated with the discipline of set topology, which defines the rules for valid spatial relationships between features in a geographic data layer.
3. *Intersect.* Geoprocessing computes a geometric intersection of the input features. The resultant features or portion of features common to all layers or assigned groups of same shape type (called a feature class) will be written to the output.
4. *Union.* Like intersect, union computes a geometric intersection of the input features. All features with the overlapping attributes from the input features will be written to the output feature class.
5. *Extraction (clip, query).* Like overlay, extraction is also intimately associated with the discipline of set topology. Queries help select the geographic data to be clipped or extracted, subject to a specific group of topology rules.
6. *Proximity (buffer).* Proximity is initiated through a query that selects geographic features based on their distance or proximity from other features. Geographic features include lines, points, and polygons.
7. *Management (copy, create).* GIS data management software is generally designed to facilitate the organization of a user's unique personal catalog (or collection) of geographic data. The intrinsic forms of all types of geographic data are accommodated by these applications.

8. *Transformation.* Typically in GIS, the term "transformation" means a spatial transformation, such as a datum transformation or reprojection. Geoprocessing, however, introduces transformations of different types, such as temporal or geometric transformations. This facet of flexibility finds favor among integrated user interfaces, complementing comprehensive spatial analysis within a geoprocessing software environment. Chapters 7 and 8 describe datum transformations in more detail.

In summary, the integration of geospatial data and geoprocessing interfaces into desktop computing makes possible the widespread use of interoperable geoprocessing software. Geoprocessing enables a clear pathway for geospatial data products in the information infrastructure. Geoprocessing opens the doorway to more defined data examination, primarily spatial analysis.

Spatial Analysis

By their very nature, geographic data are intimately related to locations and feature attributes. Spatial analysis harnesses this duplicity through the study of geographic feature locations and shapes. Spatial analysis offers the user a range of procedures, tools, and interfaces varying in application and complexity. For example, creating a simple map in a GIS environment is a basic form of spatial analysis. Spatial analysis is not necessarily complex, but it is a process of reducing complex relationships to something simpler, possibly bringing to attention things that otherwise would have remained hidden to the user.

Spatial analysis relies heavily on the first and most fundamental law of geography: Things closer together in space tend to be more alike than things that are far apart. This law is based upon Waldo Tobler's work, which essentially states that everything is related to everything else, but near things are more related than distant things. GIS does not wholly support this almost philosophical supposition, though it does acknowledge that near things are more related than distant things. In fact, geographic data contain spatial autocorrelation, which is the formal property that measures the degree to which near and distant things are related. GIS functions on spatial data to which it can be supposed that spatial autocorrelation is in essence the first law of spatial analysis.

Positive spatial autocorrelation occurs when features that are similar in location are also similar in attributes. Negative spatial autocorrelation occurs when features that are close together in space are dissimilar in attributes. Zero autocorrelation occurs when attributes are independent of location.

Spatial analysis, be it autocorrelation, overlay, or surface analysis, is an advanced and flexible form of geographic data analysis. A GIS offers the user a platform for specific application analysis of location and feature attributes. The end result is highly usable and application-specific geographic information. Let us look at the final product: geographic information.

Geographic Information

As mentioned previously, the flow of information begins with a collection of geographic data and through various functions can smartly coalesce in the final geographic information product. This final information product is derived and interactive and offers the user a host of capability, organization, and material for analysis. Geographic information can facilitate the analysis of patterns within nature and human societies, can affect decisions at a low level of political influence, and can even return accurate and subtle answers to compelling questions. The geographic information output obtained from a GIS is virtually boundless, limited only by the adeptness of the user and availability of raw geographic data.

Actual geographic information products vary in form and appearance. They are characterized in three distinct groups: software environments, maps, and documents.

Software environments tend to be the most vital and characteristic of distributed geographic information systems. These environments are usually simple front-end interfaces presented in a dialog-based (or oftentimes wizard-based) package. They facilitate the automation of spatially enabled information management, usually alleviating the dependence upon repetitive tasks in the solution of a problem.

For instance, a shoreline-permitting GIS application can help municipal government administrative personnel make basic decisions concerning the approval of permits for new development on shore land. The administrative employee can enter the required basic information about a particular subject parcel (usually required information on a permit application), launch a dialog-based wizard to organize the input data, and then return an analysis of the site in relationship to natural, sensitive, or protected resources potentially impacted by the proposed construction.

A purer GIS product is much less restricted by process. Some outrageously inventive and complex GIS software environments offer geographic information products in a constant stream. This ultimately poses the convenience and challenge of interactive research for decision making powered by GIS functions. However, in the case of the environmental permitting application, complexity is not necessary and simple output from the system is sufficient. Environmental permitting at any level usually becomes such a contentious process that actual physical maps are required to provide graphic depiction and substantiation of research findings and assertions. The "map" is therefore the most obvious example of a geographic information product, and satisfies the desire of the public at large to have everything presented in universally recognizable documents. *Maps* are the second facet of geographic information.

Maps represent the surface of the Earth—a surface with which all human beings are intimately familiar. Maps have a variegated array of uses, such as graphics in periodicals to supplement columns and articles, figures in permitting application documents, and large format prints mounted on the wall in the lobby of city hall showing planned municipal projects.

However maps, be they printed on paper or as an electronic image, are static information tools. They are noninteractive documents. GIS software environments are capable of organizing elaborate and diverse computations, occurring simultaneously with attribute or feature data. This is ideal for tabular figure information or in many instances complete report templates. This is the third facet of geographic information.

Tabular figures and other data *documents,* such as reports, charts, and graphs, are the third and final category of geographic information products. The document types help to disseminate computational, research, and spatial analysis results to anyone interested, be it the public or an engineering project manager.

These information products are usually developed and produced within a GIS environment through embedded controls native to the types of desired outputs. In some cases entire software programs for reporting, charting, Internet publishing, and database connectivity are "bundled" with a core GIS environment. This "bundled" convention encourages the user to stay right in a GIS software environment to complete tasks that are not commonly associated with GIS. Approaches like these help to expand the breadth of GIS as a modern technological discipline.

Even though geographic information is the end product of a GIS, it is not the ultimate finish line for geographic information flow. Geographic information moves beyond a GIS to the realm of limitless real-life applications. Traditional applications have been hydrologic plant relicensing efforts, infrastructure planning, shoreline zoning disputes, and wetland delineation. GISs have only just begun appearing in many nontraditional applications, such as defining statewide educational results, marketing efforts, public policy, and, yes, even presidential election result reporting. The only real stifling factor to the use of a GIS and the geographic information it offers is the user's ability to be creative, judicious, and purposeful.

2

Why Use a GIS?

In the previous chapter, we discussed to some degree how a GIS works, though we made no mention to why one would use such a system. For what reasons would someone want to progress geographic data through a transformation into geographic information? We begin this chapter with the intent to answer the question: Why use a GIS?

In 1854, Dr. John Snow did something revolutionary: He turned geographic data into geographic information and boldly stopped cholera. Let me explain. From 1831 through the mid-1850s, London, England, suffered from deadly cholera epidemics. Cholera is an acute infectious disease caused by a bacterium in the small intestines. Back in the early to mid-1800s, the disease traits of cholera were yet to be wholly identified, while outbreak sources were almost always unknown. An outbreak of the disease was a ruthless, communal disaster.

Then in 1854, a severe cholera outbreak caused horrific anguish to Londoners. In the area surrounding the intersection of Broad Street and Cambridge Street, more than 500 deaths were recorded in just 10 days. London's officials were baffled as to a solution and feared the absolute worst.

Dr. John Snow, a British physician, took matters into his own hands and mapped the cholera deaths on a hand-drawn street map of the London neighborhood. It soon became evident to Dr. Snow that a concentration of deaths centered around one particular water pump on Broad Street. He immediately presented his findings to the officials and had the pump handle removed. Immediately afterward, the spread of cholera ceased and the outbreak was finally contained.

Much like a modern-day GIS, Dr. Snow's geographic data, which included a street map and cholera death point locations, developed into a unique and highly accurate geographic information map. Without the physician's information product, countless more Londoners would have died.

GIS enables developers to take a hard look at their project goals and offers capability for developers to maintain an underlying objective to improve the overall process. Like any pervasive process, a fundamental understanding as to why each goal must be achieved is imperative to the project's general success. This fundamental understanding affords the user new opportunities for use. In fact, many times after project goals are reached, additional objectives and project efficiencies arise out of the accessibility and plethora of GIS data. Dr. Snow reached his goals to preserve human life through technology, but his product and the process by which he reached his goal were reused for new outbreaks.

The following discussions highlight typical GIS products and outline several basic reasons why GIS is the forerunner in geographic intelligence. Even though the days of usable hand-drawn maps are long gone, creative information products, like Dr. Snow's creative methodology, set apart the typical GIS user from the atypical GIS thinker.

Discerning Your Need for GIS

Specific software is designed to satisfy a specific need. Tax preparation software (such as Quicken), for example, is designed to provide systematic tax form preparation, while Microsoft Word is designed to provide powerful word processing capabilities. Each software package has a definite, and separate, function, and ultimately achieves an explicit outcome. While the need for each software package is delineated and valid, the users, in the most basic sense, must first know what they want to accomplish before using the appropriate software package.

GIS is no different. A user must have an understanding of what overall outcome or expectation is desired before implementing and using geographic information systems. Discerning the particular need for a specific application early in the project is crucial to a GIS's overall success or failure. Many times a GIS fails because the project manager hastily implements a GIS for the sake of implementing it and pays little or no attention to what the system can actually do. Failure to initially define the expectations leads to a costly GIS implementation that can increase cost exponentially as the project phases progress. Consequently, a GIS's true capability and core benefits can be lost due to poor management planning.

Unlike GIS users, lawyers are trained to ask only those questions to which the answer is known. In the courtroom, this methodology proves quite efficient. However, in GIS, knowing the actual answer to a data request or the specifics of the geographic outcome is usually never possible. Therefore, GIS professionals and managers alike must generalize anticipated outcomes for the sole purpose of management planning, such as anticipating a forest classification map of a 10-acre area of rainforest in Belize or a map detailing car theft in Madrid, Spain.

More times than not, the GIS's outcome provides greater detail and a wider range of possible analysis than first anticipated. A wholly environmental-related request can present additional detail for use with agriculture, conservation, and public safety analysis. Users need to define expectations early on, but be aware that the anticipated outcome may be a variant or enhancement of such expectations. As for lawyers and other professionals who need to know the outcome up front, take heed that GIS results can be enlightening and unexpected.

Common GIS Products and Bi-Products

GIS is one of those broad-ranging technologies that seem to have virtually no bounds other than the limit of the user's creativity. GIS, by abstract function, is a theoretical "black box" that takes data input and data requests and outputs the data results. If the user is vastly creative with both the input and requests, then the output will retain a certain degree of inventiveness. What gives GIS the aura of virtually no bounds is that the user can also implement a level of creativity with the data results, which, in the end, can produce an enormously innovative product.

Let us look briefly at the data process. As we know, standard geographic data resources are typically the only forms of basemap data available and used, such as digitized maps, census data, aerial photography, and so forth. These data are relatively static and the user is commonly bound by accessible data sources. Data requests, on the other hand, are formed by a particular application need and afford the user a suitable reason for specificity. This enables a level of user distinction with requests. An impetus for a data request may be the need for wetland delineation data in a square mile section of Brazilian shoreline or concentrations of emitted chlorofluorocarbons (CFCs) throughout Lithuania. Ultimately, geographic data requests allow creativity to enter the data process.

On the other end of the black box are the geographic data results that form the GIS product. We already discussed GIS information products in general and to some appreciable length. The GIS products are where the user's originality truly comes into play. Maps can be overlaid with multiple layers illustrating important and application-specific data.

For example, the New York City Department of Parks and Recreation used GIS to chart and manage the spread of the Asian longhorned beetle (ALB), a degenerative insect that bores holes in trees to lay eggs and whose larva bore additional holes to get out of the tree. ALBs cause the quick decimation of healthy tree populations. Through nontypical agency-created beetle data and Census Bureau data, the city was able to generate an ALB infestation geospatial application that detailed infested tree density, quarantine zones, and infestation survey areas. These common, though project-specific, GIS products helped and continue to help the agency manage the ALB infestations. Overall, highly usable geographic data sets were created, which have ultimately helped contain the city's ALB infestations to just three incidents.

Another example of common GIS products used for unique scenarios involved a series of dreadful brushfires that ravished Sydney, Australia, in December 2001. Commissioned to discern the extent of the fires by the New South Wales state government, the Sydney Catchment Authority (SCA) utilized prefire and postfire imagery and developed the Satellite Pour L'Observation de la Terre (SPOT2) data set. SCA discriminated the normalized difference vegetation index (NDVI), a gauge of land cover suppleness, for both images and used GIS to calculate burn severity between the same areas in both images. Very little or no change in the NDVI value meant that the area was unaffected by the brushfires. A large change in the NDVI value indicated that the brushfires drastically burned the area. GIS also detailed a scale of varying degrees of forest burning, depicted on change analysis maps with a colored scale. Overall, the SCA's use of GIS not only helped detect the overall extent of the brushfires but also armed the SCA with a better grasp of the nature of similar brushfires.

Common GIS products such as the density maps and graphical visualizations are typical staples for user analysis but can remain distinct through the unique representation of data results. Depending upon the user's pioneering and innovative character, the common GIS information products are not the end of the data process. Many GIS information products possess what are called "bi-products," or developable information products formed as a result of a specific application data process.

Simply stated, geographic information bi-products are potentially exploitable proprietary sets or procedures that come out of a specific GIS process. The New York City Department of Parks and Recreation, for example, was left with a one-of-a-kind geographic data set of Asian longhorned beetle data (GIS bi-product) that may be usable by or salable to some other agency or private firm. Through the brushfire GIS application, the Sydney Catchment Authority developed two bi-products: a reusable procedure for the calculation of burn severity (which is now standard practice at the SCA for brushfires), and the forest burn data results (now being used for another GIS application—Sydney's spatial fire management system).

Bi-products spawn from creative approaches to specific applications. Some unique and profitable bi-products involve geographic data sets and demographics. The process of creating special data sets is tedious, time consuming, and often extremely difficult. In some cases, it can be impossible to get the raw data. Firms are many times willing to purchase prefabricated data sets from distributors in order to minimize project development expenditures and reduce development time on the project schedule. This need has manifested in the creation and subsequent popularity of online geographic data marts/depots, where distributors compete for sales.

Specific demographics are usually harder to get and can be impossible at times without agency permissions and cooperation. Demographics add substance to a geographic data set and offer the user an enhanced capacity for data analysis. GIS developers know that qualified demographics, especially if you have proprietary access to such data, can be an enormous moneymaking bi-product from a GIS process. In fact, many firms now specialize in one-of-a-kind data sets and demographics, because of their high demand and financial value.

Database Management Systems

This section goes to great lengths to explain *how* database management systems (DBMSs) operate, but in the end the section clearly answers *why* DBMSs should be used and the functional role they play in a GIS. DBMS elements, such as queries and attributes, are covered in the previous chapter, since they are vital components to the overall flow of information but only a fractional part of a GIS's inner working.

A database management system is a program or collection of programs that enables the user to save, modify, classify, select, and extract information from a central database. DBMSs are used throughout the world and throughout almost every industry for the purposes of centralizing information and automating processes. For example, automatic teller machines (ATMs) function by means of a database management system. The user's card number and personal identification number (PIN) are entered into the ATM (database log-in security), information is displayed (database queries), user action is taken (information is sent to the central database, calculations are completed, and the database is appended with the new information), ATM dispenses money (database condition is met and approves machine's mechanical functions), and the user quits the ATM (database is closed, database log-in security activates again).

A typical DBMS is an organized, noninferential, and nonreferential data system that utilizes a file tree structure or file hierarchy. DBMSs are great for single purpose usage, where there are few users at one time and a modest amount of existing data at any one time. Without spatial coordinates, DBMSs can be described as information systems that are used for nongeographical databases (such as in the previous example of the ATM).

Ultimately, DBMSs are tremendous data management tools for small- to moderate-sized applications. These systems handle data rather easily and without much delay. However, problems arise in larger databases. Due to the amount of data, large databases not only form object-oriented redundancies, they also slow down the query processes and become a substantial programming maintenance liability.

Back in the 1960s, these database problems were already evident. Surely you have heard the stories about the first mainframes and their physical bulk. Some of the largest corporations had buildings filled with room-sized computers that had limited processing capabilities and were getting "tied-up" with tedious database query functions. Databases were highly data dependent and, as a result, the functionality was impaired when large groups of records were accessed. The most prominent and baneful forms of data dependence included ordering dependence, indexing dependence, and access path dependence. The unanimous conclusion was the urgent need for less programming, simplified functions, and compressed data files.

In 1970, Dr. Edgar F. Codd offered a valid solution with his remarkable paper entitled, "A Relational Model of Data for Large Shared Data Banks." Dr. Codd's paper detailed a new method of simplifying database management using relational algebra for simple database functions. Codd's relational model broke new ground for eliminating redundancy and simplifying database functionality through user-defined relationships between data.

At first, Dr. Codd's paper went ignored until developers at International Business Machines (IBM) took notice. Based upon Dr. Codd's groundbreaking idea, IBM developed the first relational database called System/R. As the backbone to System/R, IBM developed and implemented a new form of relational programming language called Structured Query Language or SQL (pronounced "Sequel"). The space- and time-saving element of SQL was that it defined what data operations were used rather than how the operations were performed. As a result, the amount of programming code was drastically reduced through SQL, with many cases being reduced by more than half.

This leads us to the advent of the relational database management system (RDBMS), which is a program or collection of programs that enable the user to build data relationships to efficiently save, modify, classify, select, and extract information from a central relational database. RDBMSs are predominately SQL and Visual Basic-based with some of today's most widely used systems being Oracle, Microsoft SQL Server, Informix, DB2, and Sybase.

Relational databases are typically tabular data with a series of data columns and rows. The columns or *attributes* contain distinct field data, while the rows or *tuples* contain the individual records. The relational aspects of these databases are wholly contained within the attributes. In addition, each record has a unique attribute called a *primary key*, which distinguishes that record. These primary keys can be sequential record numbers or user-defined values that are unique and, thus, not duplicated anywhere within the same attribute throughout the records.

Consider Figure 2.1. Two distinct tables exist with various attributes and records. The relationship between both tables lies within the "Customer" attribute. This single relationship makes available to the user a wide range of query and data selection capability. Attributes "Customer" in Table A and "Order Number" in Table B are the primary keys.

RDBMSs can take on a variety of system architectural types, but the most fundamental and commonly used types include file-based systems, client/server systems, and multitier systems.

A *file-based system* is the simplest of relational database types and uses one file that contains the database tables, queries, security information, and forms/reports. Microsoft Access is a perfect example of a file-based system. The database file can reside on a single user computer or on a network. This type of architecture fails when more than a few dozen users access the database at one time.

A *client/server system* is designed to effectively handle many users accessing the central database at one time with speed and efficiency. The client/server system typically uses a network to connect the user workstations (called clients) to the central database residing on a file server. The system name evolves from this arrangement, hence "client/server." A client accesses the database using SQL and Visual Basic commands, and, in return, receives data from the database file server. The client, in turn, uses his workstation to process the data and perform the necessary calculations and report generation (i.e., information product). The amount of data transmitted from the file server is, in most cases, substantially less through the progressive use of SQL/Visual Basic query commands and user workstation processing.

The client/server system is a remarkable advance from the file-based database system but has one unmitigated fault: scalability. Data scalability involves the power to reduce or increase the amount of data being transmitted from the database to the user. Often, data scalability is warranted when information is presented on the Web or if data accessing rules must be employed.

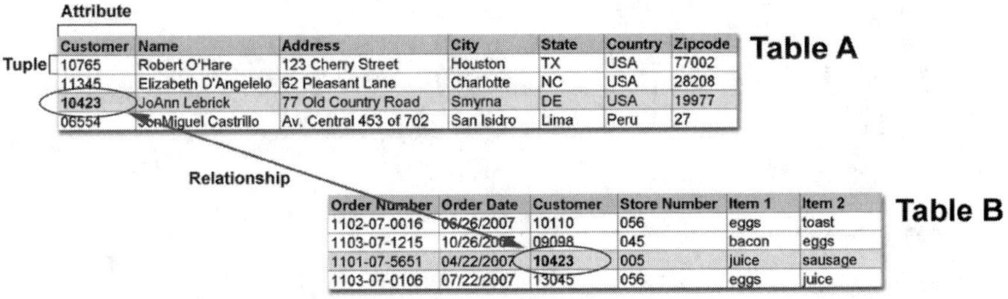

Figure 2.1 A relationship between tabular data.

The solution to scalability is a *multitier system*, whereby there are one or more data task servers between the database server and the client. A multitier system (sometimes called a three-tier system) functions somewhat like the client/server system, except that there are middle servers, which contain software to perform certain tasks and process some of the transmitted data. These middle servers help reduce the amount of data being transmitted to the client.

The use of multitier systems results in easier application deployment, data scaling for the Web, enhanced security, and the ability to implement data or business rules. Since clients have reduced processing capabilities in a multitier system, they are called *thin clients*. Conversely, clients requiring their own individual rules, such as in the client/server system, are called *thick clients*.

Figure 2.2 depicts the three fundamental RDBMS architectural types and their differentiating physical arrangements.

So now that we have discussed the physical and functional infrastructure of relational database management systems, we can ask: *Where does GIS fit in?*

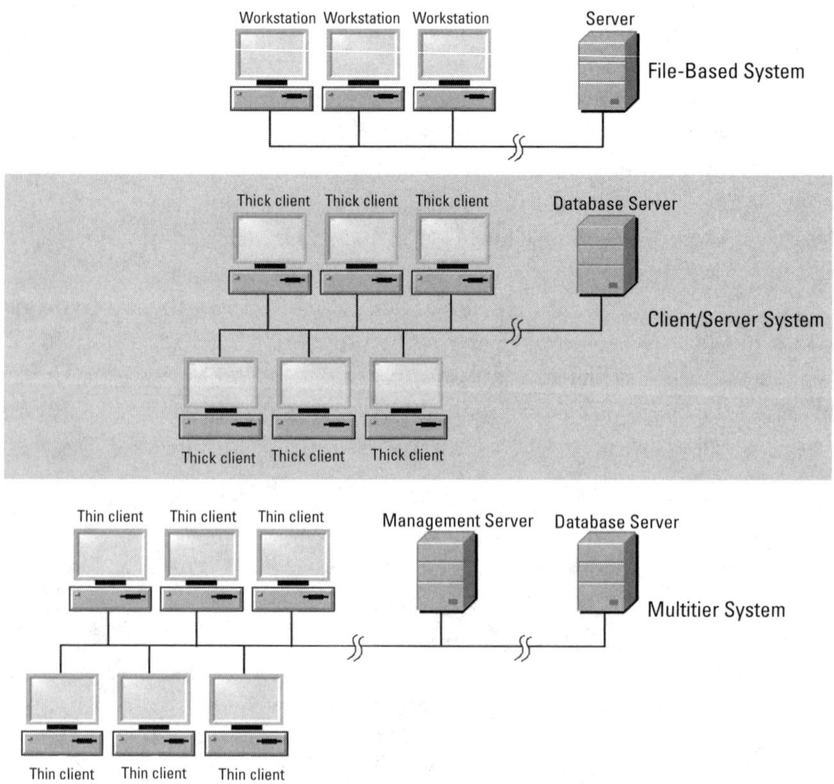

Figure 2.2 Three fundamental RDBMS types.

Geographic information systems follow the multitier system architectural type in that there are typically three tiers: a public presentation tier, a business and data management tier, and the GIS management tier.

The public presentation tier involves general viewing, Web-enabled data access, superficial analysis, and geographic visualization. This first tier is made up of general end users (or public thin clients). The business and data management tier involves users responsible for building the geographic database, site maintenance, critical analysis and visualization, and project management staff. This second tier acts as a filter to what information can and will be presented to the general public, as well as creates the project-specific geodatabase. The GIS development tier involves GIS professionals and systems responsible for data modeling and design, Web site development, low-level programming, and site administration. This third tier can also be called a data tier since it primarily functions to supply the entire database information.

Figure 2.3 depicts the three typical GIS tiers and their interaction with the RDBMS. Notice that the GIS development or data tier provides a relational and

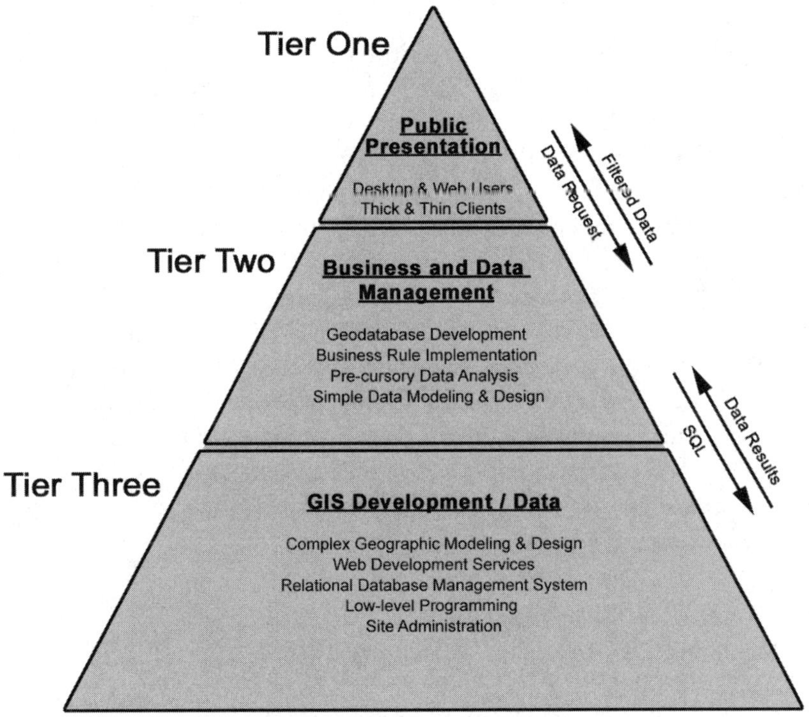

Figure 2.3 The three common GIS tiers.

comprehensive data resource. The business and data management tier takes the entire database and creates a project-specific geographic database, weeding out extraneous information and saving the pertinent. The thin clients then utilize both desktop and Web-enabled applications for viewing and analyzing this data. Overall, a massive relational database management system forms the backbone to GIS.

Let us look at a real-life example to help understand GIS's three-tier system. MapQuest.com is a free GIS-based map-generating product on the Internet, whereby a GIS backbone populates the Internet search capabilities. Although entirely different, GIS and the Internet integrate together to provide a Web-based front for geographic information. In brief, Web users have the ability to enter basic location information and view corresponding maps of that particular world location. Consider the user as the thin client in the public presentation tier (first tier). Users can view maps through minimal location input. Say, for instance, you are the thin client and you request a map of 415 Charles Street in Omaha, Nebraska. The software sends a query for data to the business and data management tier. This second tier receives the request from the user and processes the information. The second tier in turn sends a request to the RDBMS in the data (third) tier for Omaha, Nebraska, map data. The third tier contains every map for every city in the world, scalable to street level. Once it receives the request, all Omaha, Nebraska, map information is transmitted back. The second tier receives the RDBMS information and selects all the Charles Street information. This selected information is then searched by addresses until address 415 is found. The second tier transmits this select data to the first tier software, which, in turn, creates a graphic map. The thin client now has the correct map requested through the Web.

This leads us into the topic of how databases utilize the Web. In fact, there is a delicate bidirectional relationship between the RDBMS and the Web that patterns the very nature, flexibility, and importance of GIS Web services. All RDBMSs are designed to distribute information on the Web with a varying degree of user interaction. To achieve control, RDBMS utilize three types of data distributing Web pages: static, dynamic, and active. Because GIS is a relational geographic database management system, GIS products are portrayed and distributed in the same manner.

Static pages display database-driven information that cannot be changed, altered, or modified by the user. In fact, static pages display static Hypertext Markup Language (HTML) content. Dynamic pages, on the other hand, involve some user interaction, whereby the pages display the most recent or dynamic database information as defined by the user. Dynamic pages access a database through a query. MapQuest utilizes dynamic pages. The most common implementations of dynamic pages are with Java, a multiplatform, object-oriented programming language. Active pages involve full user interaction, whereby the user can request information as well as make changes to and manage the database. Active pages can be considered a comprehensive database terminal with processing logic capabilities. In any case, HTML pages are the most common type of Web page utilized by GIS Internet map servers.

GIS Web services utilize all three Web page types when portraying geographic information. Noninteractive map products would typically be displayed using static pages, since user input is not required. GIS professionals use dynamic pages when the end user requires some functionality, such as turning on and off layers to a GIS product. The United States Environmental Protection Agency (http://www.epa.gov) utilizes dynamic pages so that users not only can search regions for reported environmental hazards but can customize the views through interactive layer selection. Lastly, GISs use active pages so that GIS developers have full control of the background data.

In conclusion, database management systems, both object oriented and relational, are important not only to how a GIS works but, more specifically, to why GIS is used rather than other methods. A geodatabase, for instance, is implemented directly on a RDBMS to take advantage of the system's data management capabilities and system administration tools. In turn, a GIS expands the functionality of a relational database so that it can proficiently handle geographic data, produce custom maps, and perform spatial analytic tasks. Together, GIS and a RDBMS offer the best in database management technology, geographic data distribution, information manipulation, and accelerated processing.

Quality Assurance, Six Sigma, and GIS

A fully implemented GIS adds value to a project while maximizing accuracy and data quality. In today's industries, quality is a big subject. You cannot enter any industry these days, let alone the engineering and environmental fields, whereby quality is not a major factor as to whether you will succeed or fail. In recent years, phrases such as "ISO-certified" (for International Organization for Standardization) and "ISO-9001 standards" are commonplace and, in many cases, required to provide services for certain agencies.

By its very structure, GIS elicits a high degree of data consistency and source data reliability. GIS information products, such as layered maps, contoured reliefs, and point data, offer various automated procedures and same-source data. GIS's process automation is just one placard for achievable higher quality output.

A rising new methodology for achieving quality that is quickly gaining corporate placement is called Six Sigma. Six Sigma stands for six standard deviations from the mean or average and involves a measure of quality that strives for near perfection. The lowercase Greek letter sigma is the mathematical representation for standard deviation. In practice, Six Sigma is a disciplined, data-driven methodology for eliminating shortcomings and deficiencies in any business process. This perceptive approach can be implemented from the development stage to the service provision stage and involves staff trained in the Six Sigma principles. Large companies, such as GE and Motorola, already implement the Six Sigma approach and have experienced the benefits.

Unlike the ISO or ANSI (American National Standards Institute) standards, Six Sigma is not a regulation or a directive, per se. It is simply a unique way to achieve higher quality and save money. Incidentally, GIS can offer some of the same quality-enhancing results. GIS is complementary to the Six Sigma principles through GIS's data-driven structure, invariable consistency, and ability to improve processes. In addition, like Six Sigma, GIS's capability to reduce data variation and efficiently improve an organization's data quality is quickly becoming the new industry benchmark.

Developers, managers, and users are seeing first hand the far-reaching advantages to implementing an enterprise-wide GIS, from project efficiencies to fiscal brow beating. Its heightened quality enhancements complement existing corporate quality programs. Six Sigma and GIS can become the one-two punch for quality preeminence and knockout ISO/ANSI regulation compliance.

Project Visibility and Popularity

Through fluid Internet accessibility and high-octane map controls, GIS has become the preeminence in data sharing and distribution. Many times local governments and agencies find geographic information systems to be the one-size solution for their communication and public outreach programs. By default, GIS has become the technology of choice for detailing spatially based technical results to nontechnical viewers. The use of data representation and techniques makes analysis easy for those not directly involved with the project or experienced with the processes.

GIS is highly visible. Users can access data results easily through the Web and can manipulate (to some extent) how the data can be presented onscreen. For instance, after September 11, 2001, the U.S. Environmental Protection Agency (USEPA) let users view mapped environmental contamination and air quality data for Lower Manhattan. Users were able to turn on and off information layers over a street map baseline. The USEPA updated the air sample results daily and millions of users accessed the site. Through necessity and huge public interest, this basic GIS-based site experienced incredible visibility and popularity.

On a more fundamental and everyday level, project managers have embraced GIS use because they can efficiently manage the project's document and graphic-based results from a single data location. Through GIS, virtually limitless project personnel and management alike can review the results in real time via the Web. The increase in project communication and quick project team debriefing has made GIS highly popular in project implementation and administration schemas.

As another example of GIS's popularity and visibility, a soil sample for a transportation rehabilitation project can be taken in the morning and its location, analysis type, and other relevant data can be easily entered into the GIS. Soil sample data locations and base information is instantly accessible by team members and others with GIS access. Soil sample information can be illustrated on a project site map with hyperlinks to soil boring logs and, eventually, the laboratory results. In the end, the benefits of GIS's one-stop shop of project data make GIS an extremely popular and highly exploited project tool.

Benefits, Cost Savings, and Automation

In many businesses where the bottom line drives decisions, GIS is often viewed as an expensive resource. Most decision-making managers have only the vaguest idea of what a fully enabled, successfully implemented GIS can do for their business proceedings. Although the concept of implementing a GIS is typically interesting, the front-heavy costs of implementing the system are not. These initial GIS costs are often enough to scare away most managers, while the proposed automated and cost-saving outcomes get classified as a project whim.

Today GIS is breaking new ground in project efficiencies, data communication, and value-added deliverables. Many private firms and public agencies are now requesting the implementation of a GIS on their projects because of the overwhelming long-term savings and advantages. Especially in local governments, GIS has made great strides to help consolidate the vast types of data and information into one categorical data repository. Local governments have begun learning the wide-ranging savings involved with distributed data throughout their projects. The dreaded fiscal black hole of government data communication is becoming a thing of the past.

One such local government is the Maryland–National Capital Park and Planning Commission (M-NCPPC). In a cost-benefit assessment report from 1999, M-NCPPC detailed the overall impact of the GIS implemented in Montgomery County, Maryland, highlighting the significant changes from the pre-GIS county environment and detailing the outstanding benefits of the GIS. Some of M-NCPPC's pre-GIS problems included inaccurate maps and data; incompatible maps and data sets; limited numbers of maps produced because of the required person-time; and end products not being standard. In total, M-NCPPC detailed nine crucial problem areas in the pre-GIS environment.

Since Montgomery County implemented GIS, a number of benefits were achieved. These benefits were grouped them into five areas: (1) overall improvements in existing operations from GIS, (2) additional capabilities not available in a non-GIS environment, (3) emergency response using GIS, (4) intangible improvements, and (5) product and bi-product sales revenues.

M-NCPPC's end result in its cost-benefit assessment report was that Montgomery County saved a typical annual amount of more than $500,000 from GIS functions alone. Additionally, GIS bi-product sales from fiscal year 1999 amounted to well over $44,000. M-NCPPC's overall conclusion was that the use of GIS had an overwhelming functional and cost benefit to Montgomery County.

Local governments have also experienced the benefit of increased government efficiency with the implementation of a GIS. For instance, the city of Ontario, California, used GIS to audit their billing files and reclaimed $190,000 per year in lost business license fees. In Bucharest, the Romanian National Railways Company implemented an ambitious GIS to automate their railway systems, which ultimately resulted in a geographic inventory of all railway assets, geographic maps for more than 22,000 km of track, and a centralized database to support all primary business unit applications. Additionally, a European Commission–backed project called Bacchus is using European Space Agency imagery and GIS to chart Europe's vineyards in unprecedented detail. The GIS information products provide vine growers with the tools to improve grape production, manage vines, and guarantee grape quality.

Just looking at the enormous benefits and heightened automation within the handful of examples presented, the intrinsic power and enormous impact that GIS has worldwide is easily understood. The key to maximizing all GIS benefits and cost savings revolves around knowing what the user wants from the geospatial application. The earlier the users can define their expectations, the cheaper the GIS implementation. Incidentally, user education is one sure-fire way to secure continued benefits from any geospatial application and helps the untrained user eye become more discriminating from the project start.

Possibly the greatest benefit of GIS is the ability to conduct tasks not previously possible. Whether it is Europe's vineyards or Maryland's local government, GIS has made an indelible mark on the procedures, efficiency, and future health of public and private entities worldwide. If you need additional proof as to the overall success of GIS and more reasons why GIS should be implemented, then peruse the voluminous case studies and success stories available on ESRI.com, Intergraph.com, and the hundreds of local government sites throughout the Web. You will soon recognize that the question *Why use GIS?* has but one accurate response: *Why not?*

3

The Structure of Geographic Data

Geographic data come from a variety of sources, such as digitized maps, aerial photography, GPS, and field data. Valid geographic data serve as a collective true marker as to what level of detail and accuracy a GIS can potentially attain. But, in reality, this is only a half-truth. The manipulation of this geographic data is as important and, often, more suggestive of the outcome's quality.

The appropriate structure of geographic data is considered a top reason why a GIS succeeds or fails. Inappropriate data handling or structure predictably leads to inappropriate geographic information products. To this extent, a user's grasp of the GIS application, the desired outcomes, and the geospatial system's limits are detrimental to realizing how best to treat the geographic data. Ultimately, the successful structuring of geographic data requires a combination of understanding and sound decision making.

Geographic information systems utilize two primary data models to manipulate and structure geographic data: the raster data model and the vector data model. We have already discussed both types briefly in Chapter 1 but now must build upon that initial introduction to gain a better understanding of both the elements and idiosyncratic nature of each type. By examining the nature of each data model you will have a much clearer picture of the important and institutional structure of geographic data, as well as a distinct understanding of the advantages and disadvantages of each.

Raster and Vector Data Structures

Like any complex data structure, raster and vector data have a myriad of different realizations that vary in complexity through use, appearance, format, and file size.

Although distinctly different, these affiliate data structures share two characteristics: (1) They visually represent real-world features, and (2) they are subject to orientation within the real world. By satisfying both of these characteristics, geographic data are born and made interoperable with other geographic data sources within GIS. Before proceeding further with this discussion, let 'us take a small step back to reacquaint ourselves with the raster and vector data structures.

Raster data structures characterize continuous data (such as imagery) and are exceptionally strong where boundaries and point information are not well defined. Raster data provide data as a pixel grid, whereby each pixel or cell is a feature capable of retaining properties and attributes. These pixels approximate pictures and images in an impressionistic way, with all of the smallish, monothematic cells contributing to a greater whole. Adding further identity, a raster image can vary in file format, color representation, resolution (size of pixels/number of pixels per set area), and potential properties.

Vector data are a bit different. Vector data structures characterize discrete data (such as roads, pipelines and topographic features) and are exceptionally strong where distinct boundaries and point information are well defined. This data structure is constructed on ordered two- and three-dimensional coordinates ([*x,y*] and [*x,y,z*], respectively). Features are represented as geometric shapes defined through single or grouped coordinates on a set grid.

For a clearer understanding of both models, refer to Figure 3.1, which illustrates the visible differences between raster and vector data, as well as how each data set represents Earth features. The original graphic is a raster, high resolution satellite image of a land parcel with infrastructure and a pond. Notice the distinct feature variance between its continuous data image and vector's point representation. It is clear that the raster image serves as a good overall representation, while the vector representation serves as a good individual feature representation.

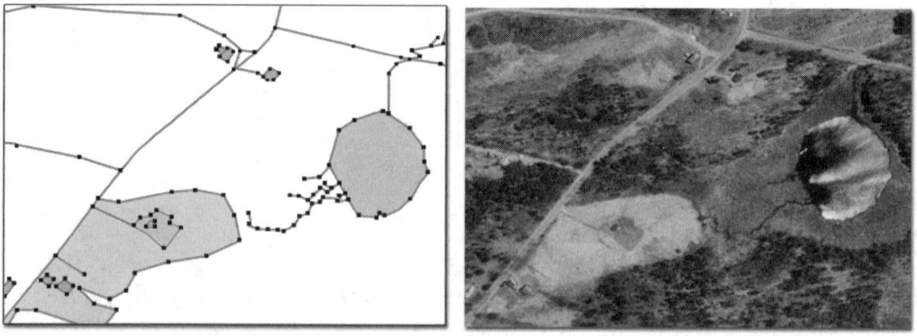

Vector Representation Raster Satellite Image

Figure 3.1 Raster versus vector data representation.

Both the raster and vector data structures have inherent advantages and disadvantages that allow GIS users a certain degree of choice. A full understanding of the native characteristics of each data model is a prerequisite for GIS success. Often certain requirements force the use of one data model rather than the other, such as the need for a better output resolution, easy image analysis, or enhanced spatial accuracy.

Raster data, for example, offer a truly simple data structure that involves a grid of row and column data. This simple grid structure allows for easy raster image analysis, as well as analysis among multiple images. Raster modeling is also much easier to implement due to the single-value cell structure and relatively simple software programming.

These advantageous raster capabilities are shadowed only by raster's native weaknesses. Disadvantages to raster data include general spatial inaccuracies and misrepresentations, low resolution, and massive data sets that require significant processing capability. The lack of accurate topology is also a major raster-based limitation.

Similarly, vector data offer their own variation of modeling strengths. Vector data, for instance, are spatially accurate and support a better, higher resolution than the raster data model. The ability to provide topology or feature relationships is a definite advantage, as well as the minimal data storage requirements.

Although seemingly ideal, vector data have their own share of weaknesses worthy of mention. Due to the complex data structure, vector data require a greater and more powerful processing capability. With this comes the need for better, faster workstations to minimize the data processing times that typical computers face. Inevitably, the costs to run a vector GIS and expeditiously process complex geographic data sets can become highly expensive.

Needless to say, both the raster and vector data models have their own brand of geographic data expertise and capability embedded within a GIS. Nowadays, GISs are offered as raster-based systems, vector-based systems, or raster-vector capable systems. When necessary, raster-to-vector/vector-to-raster conversions can be easily accomplished through GIS or third-party conversion software. Undoubtedly, the choice of data model lies with the user and the available software/hardware.

The next few sections examine raster and vector data capabilities that are vastly used in everyday GIS work.

Vector Feature Geometry

Often it is necessary to depict real objects as features on a map and designate object positions within a GIS. These features can range from the simplistic (i.e., linear transmission lines) to the complex (a multibranched, nonlinear river). To detail objects as features in a GIS, we must choose the best data model. In most cases, features are defined using the vector data model.

Given the nature of raster data, features are ambiguous in shape and general in position. For applications where specific form or position is not needed, raster may prove easiest. Some possible applications where raster data can be used to define features is on nonprecise maps, such as visitor maps of an amusement park, or macroscopic markers on an overview map, such as a colored box for approximate position.

Alternately, given the accurate, positional nature of vector data, features are best represented by coordinates and geometry. Real-world objects can be represented as individual or a group of geometric shapes called feature geometries. In any geospatial platform, there are three primary types of feature geometries: points, lines, and polygons. As a subset of these three primary types there exists a fourth geometric feature called a polyline. Figure 3.2 illustrates these major feature geometries in both the raster and vector data models. The differences between and usefulness of the models are obvious as they relate to depicting feature geometry.

A point is an individual position defined as a vector *x-y-z* coordinate or as one raster pixel. Lines are two connecting points with two distinct coordinates in vector or a linear block of pixels in the raster model. Polygons are a grouping of vector coordinates connected in a sequential fashion or a group of pixels

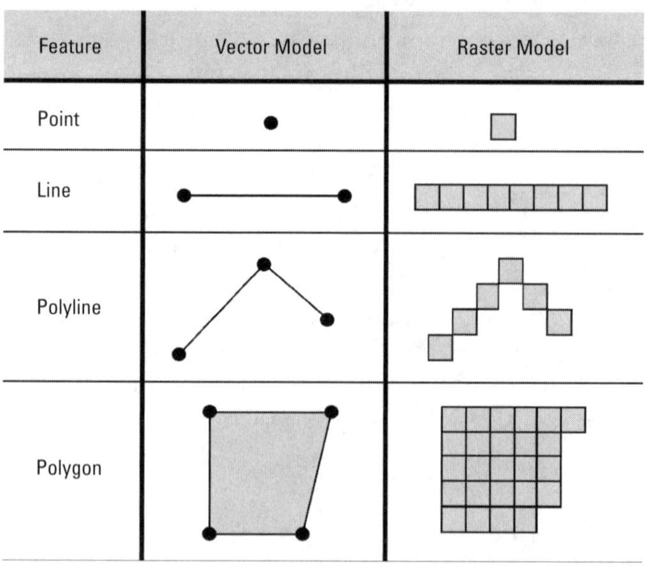

Figure 3.2 Feature geometries in raster and vector.

forming the object's general shape. Lastly, the subset polyline (known also as an arc) is a connected string of vector points or raster pixels. Often a polyline involves two lines sharing a same point or pixel (vertex).

Vector features are constructed on ordered pairs of vertices, whereby these ordered pairs reside in the design plane having known *x*, *y*, *z* locations. The positions of these ordered pairs are recorded in the vector file and often encoded as binary. These vertices are objects and can be grouped to form features that exist as independent entities. This is in contrast to the raster model, where the entire image is an object. In raster, features must first be sampled and then represented as image pixels, which, as we already know, visually approximate the features. Raster images have no independent vector ordered pairs, the actual building blocks of feature-driven spatial information.

Vector data for feature geometry comprise positional coordinates and, for some, other position-defining data, such as inside/outside and left/right. Figure 3.3 depicts the three primary types of geospatial feature geometries, as well as the polyline and the more complex polygon within a polygon (i.e., the "doughnut").

As illustrated in the figure, the point depicts a discreet position in space, defined by an individual geospatial coordinate. The line is defined by two geospatial coordinates. The open-ended third feature is a polyline, which comprises two vector lines sharing a common point (vertex). Interesting enough, polylines are defined not only by geospatial coordinates, but also by left and

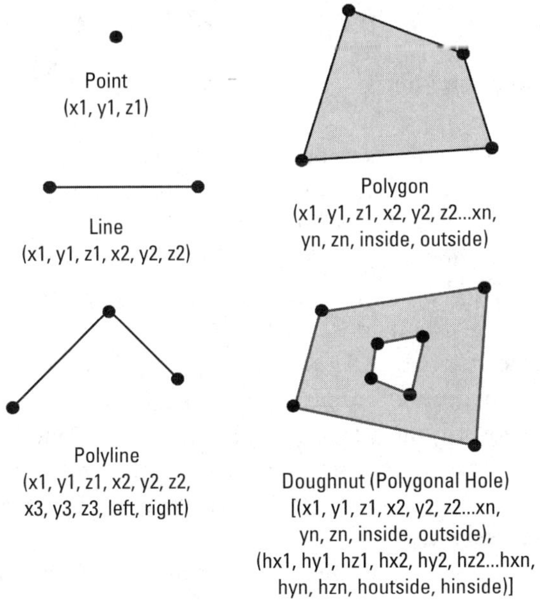

Figure 3.3 Major vector feature geometries.

right characteristics that indicate whether other features are directly left or right of the polyline.

Growing in complexity, the remaining two geospatial feature geometries in Figure 3.3 are closed features and depict specific areas. The polygon involves numerous vector points that are connected in sequence. Polygons are defined with inside and outside characteristics that delineate whether any overlapping geospatial feature geometries exist inside the polygon's boundaries.

The most complex of the geospatial feature geometries presented in Figure 3.3 is the doughnut or polygonal hole. A doughnut comprises a polygon within another polygon. Polygon 1 forms the boundary area and polygon 2 serves as the "cutout" within polygon 1. Think of a doughnut, whereby polygon 1 is the outermost doughnut boundary and polygon 2 is the cutout in the doughnut center (doughnut hole). Each polygon enables coordinate characteristics for inside and outside elements.

To uniformly structure, manage, and manipulate feature geometries like the ones just discussed, a GIS stores vector data in geospatial feature file formats. Common feature file formats, such as the *shapefile* developed by Environmental Systems Research Institute (ESRI) or the *TAB file* developed by MapInfo, are basic geometric containers compatible with an overwhelming majority of GISs. These geographic dataset types store nontopological vector (or coordinate) geometry in the form of real-world spatial features, as well as links to attribute information for these respective objects.

Raster Image Structures

The nature of images, such as aerial photographs and base maps, involves a continuous array of data. As discussed, the raster model provides the best solution for continuous data. Due to the lack of specified feature boundaries, point locations, and multiple objects, vector data are not intrinsically fit to handle the image information. A vector data representation of such an image produces a complex and often massive image structure.

Unlike a vector format, the raster palette is much more desirable for applications requiring the use of photographs or digital scans. Raster imagery instantiates outrageously complex geometries quite easily with the use of attributed image pixels. The essence and value of raster image structures are that raster represents real-world information with native visual properties—a feat vector imagery cannot reproduce.

Because raster images are physically continuous by nature, a reasonably accurate transformation process can be applied to fit raster data onto the real world. In brief, images are "rasterized" or digitally transformed to raster data through a matrix of pixels. Photographs are typically scanned with a set image resolution defined by pixels per inch (ppi), more commonly known as dots per inch (dpi).

With the use of a GIS, the organic transformation into a raster image can enable users to characterize previously nonattributed existing geographic data material. In fact, solutions produced from this process result in alternative, often unexplored, geographic data sets.

There are times when an existing raster image is not at the desired resolution for the present application and an image adjustment is necessary. Manipulating resolution within raster images is a technique to adequately control and manage the amount of raster data to be processed and the overall image quality. As mentioned earlier, raster data resolution is the result of the amount of row and column pixels used to define an image. Raster image dataset size is directly proportional to the amount of raster pixels used to define an image. The rule of thumb is: The more pixels in the grid, the higher the image resolution, quality, and dataset size; the fewer pixels in the grid area, the lower the image resolution, quality, and dataset size.

Modifying the image's resolution presents varying results within the image. You can easily transform a high resolution raster into a low resolution raster without desecrating image quality. The image size often remains unchanged through an intelligent interpolation that transforms a grouping of pixels into one pixel. However, it is not always easy to reverse this transformation. Taking a low resolution into a high resolution form is not always feasible. To retain moderate image quality, a high resolution raster will take on a reduced, fractional size to that of the original low resolution raster. If image size remains unchanged, the new high resolution image will be relatively unusable and indistinguishable in content. Care should always be taken when modifying the resolution of raster images.

In actuality, geographic information obtained from these types of organic processes may be of critical relevance to future projects and unwittingly provide solutions to problems not yet known. Many times these unforeseen criticalities involve complications put forth by the presence of precious natural resources on a project. The manifestation (or realization) of these transformations is the fundamental building block for geospatial systems and makes available opportunity for other advanced GIS techniques.

Topology

It has already been established that in a vector-based GIS, primary geometries (i.e., points, lines, and polygons) represent real-world features. Topology is the set of rules through which a GIS represents features with the primary geometric shapes (i.e., point, line, and polygon). The vector data model utilizes topology to organize spatial relationships between discrete features. In essence, the main functions of topology are to define: (1) feature-to-feature locality or, simply, where a feature is in relation to another feature, (2) what is shared between different features, and (3) how features are grouped or connected within a set.

In a GIS, topology establishes geometric harmony within a geographic data set. Illustrating this precept is ESRI's *shapefile*, in which the vector feature file employs a set of natural numbers plotted and structured in binary format. This set of natural numbers delineates the boundaries and coverage of a particular feature. The simple binary structure defines a topological space for the feature and establishes continuity with other features, forming topological relationships.

Topological relationships are defined in all types of feature files and are generally categorized into the three primary functions of topology (previously mentioned):

1. Feature-to-feature locality, called a *complement;*
2. What different features share, called an *intersection;*
3. How features are grouped, called a *union.*

In Figure 3.4, the simplistic Venn diagram depicting two overlapping, amorphous shapes highlights the base concepts of the three feature topology functions.

Within this figure, the lighter areas are considered *complementary* shapes (*A* and *B*), while the darkest area depicts the *intersection* of the two shapes (*C*).

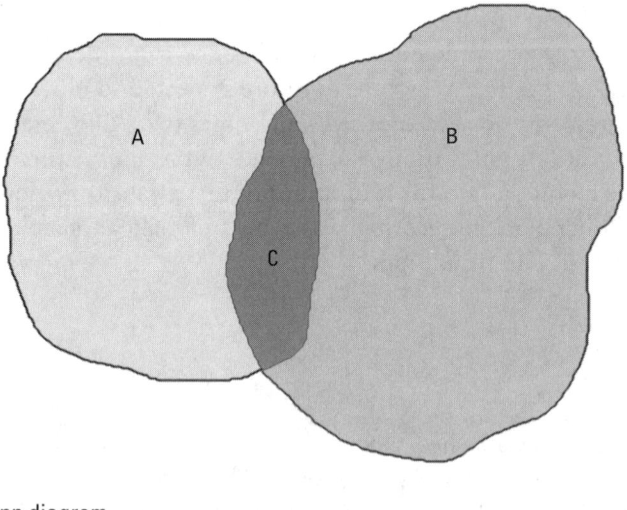

Figure 3.4 Venn diagram.

All three areas together (*A*, *B*, and *C*) form the *union* of the set. Applying these concepts to a vector GIS, shapes *A* and *B* can be considered two separate features that share topologic space (*C*).

Feature topology is also very sensitive to retaining the original shape of different features and the demarcation of each individual feature. In the context of traditional mathematics, topology is the examination of objects and groupings of objects that exhibit a predictable structure. Put simply, topology involves object aspects that remain unchanged even when the object itself is under some form of physical transformation (i.e., intersection). Similarly within the figure, shape *A* maintains its form even when sharing topological space with shape *B*.

All types of geographic data sets can utilize topology within geospatial software environments. These predictable structures (such as shapes *A* and *B*) constitute a set of features (*N*), which represents whole geospatial features. Topology implements a successor function: $s(x) = x + 1$, which succinctly indexes each separate feature for identification (FID). The successor function enables unique identification of each feature within the set. In this way topology serves as a geographic data quality overseer by ensuring the geometric integrity of geographic features between the real world and GIS, as well as retaining responsibility for producing truly clean and representative geographic data products.

Topology also helps to avoid repeating feature data, such as shared boundaries and shared nodes (points). The data model stores a single line to represent a boundary, as opposed to two lines with the same coordinates. This topological quality control helps maintain a smaller data set and vector feature file.

In Figure 3.5, polygon 1 (points *QRST*) and polygon 2 (points *RUVS*) are side by side, connected by a shared border (*RS*). Topology dictates that line *RS* for polygon 1 and line *RS* for polygon 2 are identical and the data model only accounts for *RS* once. Feature topology proves diligent in avoiding topological overlap.

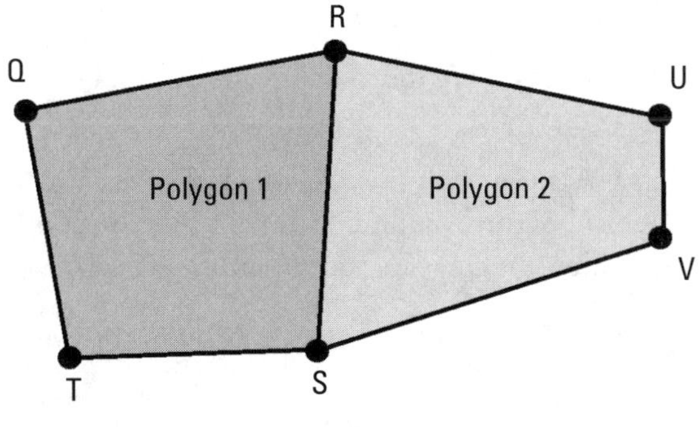

Figure 3.5 Topology of two polygons.

These geometric monitoring techniques allow GIS to control, query, and edit the topological coincidence between geospatial features (objects). Topology introduces the notion of absolute feature continuity, opening the door for numerous potential software compatibilities, including complex mathematics and engineering programs.

GIS Attribute Tables and Indices

In order for a GIS to successfully manipulate and portray geographic data, feature files must be indexed to attributes providing more information. As detailed in Chapter 1, most standard GIS data formats consist of a feature file, an index file, and a linked attribute table, whereby the feature file contains geographic object feature information, the attribute table (file) contains explicative attributes for spatial features, and the index file links attributes with features.

Briefly stated, index files act as attribute pointers. Indices contain unique identifiers that contain more detailed information about a specific feature. This embedded linkage helps speed up spatial feature queries within the GIS and serves as an index file's only true function.

Geospatial attribute tables, on the other hand, drive the spatially enabled database. There are various attribute field data types to handle the multitude of data, differentiated by a specific form of data and the degree of precision. To adequately handle the variants, attribute field data types signify different groupings of data bits, such as incorporating sign (i.e., "+" or "−"), binary, mantissa, and exponent bits, or signify data values, such as text, date, and object ID values. For clarity of this discussion, an *exponent* defines repeated multiplication (i.e., in $2^3 = 2 \times 2 \times 2$, the exponent is 3) and a *mantissa* is the value to the right of the decimal point in a common logarithm (i.e., in log 196 = 2.2923, the mantissa is 0.2923).

The following attribute field data types are most common and are supported in many major GIS environments:

- *Short Integer.* A basic attribute data type that includes one signed bit and 15 binary bits.
- *Long Integer.* A more complex form of the basic attribute type that incorporates one signed bit and 31 binary bits. As you can imagine, the *Long Integer* offers greater precision than the *Short Integer.*
- *Float.* Contains one signed bit, seven exponent bits, and 24 mantissa bits.
- *Double.* A more complex form of the *Float* attribute type with one sign bit, seven exponent bits, and 56 mantissa bits. As with the *Long Integer,*

the *Double* attribute type holds greater precision than the *Float* attribute type.

- *Text.* Contains varying forms of data, such as numbers, letters, and symbols. The *Text* attribute type is a character string that can hold any amount of characters, but each character is stored using eight bits (called a byte). An interesting aspect to *Text* attribute data are that each text value in the same field must have the same number of characters. To achieve this, end blanks are used to fill in the empty slots.
- *Date.* Though not apparent from the attribute data type name, a *Date* type contains date and time data. The value is based upon a standard time format and is automatically transformed into the current day and time within the system's local time zone.
- *BLOB.* Short for *Binary Large Object.* A *BLOB* is a complex (and large) object stored in the database that may include an image, sound, video, or geometry. *BLOBs* allow users the ability to insert any type of multimedia data into the geodatabase.
- *GUID.* Acronym for *Globally Unique Identifier.* A *GUID* is a unique 128-bit (16 byte) number that is produced to identify a particular application, file, database entry, hardware, or user. Each generated *GUID* is "mathematically guaranteed" to be unique since the total number of unique keys is colossal and the probability of generating an identical *GUID* is virtually impossible.

As with any abstract concept, casual or novice users often misunderstand attributes in geospatial systems. With a better understanding of GIS's theory (this book) and real-world practice, these feature-attribute concepts will congeal.

4

Geospatial Metadata

Today's society is information rich. We can walk into any major bookstore and use a computer to locate individual or related topic books, obtain in-stock status, and get exact store location. Let us not forget we will probably be provided a host of other miscellaneous information about the book, such as its ISBN, the number of pages, and reviews. In a matter of a few minutes, we could identify whether or not our visit is fruitless. In that short time span, we not only have discovered whether our desired information is available but also have saved a great deal of time.

This act of information sharing and dissemination is, by all means, not by accident or some form of electronic wizardry. It is a well-organized data retrieval system that retains information in a real-time database. The search function utilizes the searcher's entered information and locates specific record fields or keywords with any associated matches.

Similar to the bookstore's search utility, data resources, such as data sets, images, and documents, are digitally marked with descriptive tags known as metadata and are organized in a specific, universal order. Searches can be made against the metadata to locate these resources quickly. Without metadata description or encoding, each resource must be searched in its entirety for a match. You can just imagine how long it would take to search millions of individual resources in their entirety.

This chapter attempts to take the wizardry and mystery out of metadata. It offers an overview of dataset metadata and demonstrates how a defined metadata structure can serve as a key time-saving protocol for GIS users. It details some of the efforts taken to organize and expand available resource data, while presenting the geospatial paths taken by worldwide governments. Due to the nature of this book and the wide breadth of metadata, this chapter is limited to an overview. Specific field metadata and metadata authoring (see Table 4.1 later in this chapter) are subjects for further exploration.

What Is Metadata?

Metadata is defined as structured information that enables a resource to be easily identified, used, manipulated, and cataloged. Metadata is often identified as "data about data" and can be applied to various resources, such as images, online documents, maps, library records, data sets, or anything else inherently searchable. In simpler terms, metadata helps resources be suitably found, cataloged, and used.

The concept of metadata is nothing new. In fact, library card catalogs, which have been around since long before GIS, are a form of metadata. The older hardcopy information cards and newer online records hold a selection of basic information and identifiers for the material being sought. Typically, information supplied includes a title, author, date, ISBN or similar identifier, and a unique library location number. This library catalog card/record is identical to metadata since it is data (card information) about data (book or resource material). Figure 4.1 identifies some common manipulators of metadata.

Metadata sometimes is embedded within the resource and other times in a stand-alone document. For instance, Web pages have metadata embedded into their source code, often programmed in HTML, and image files have embedded

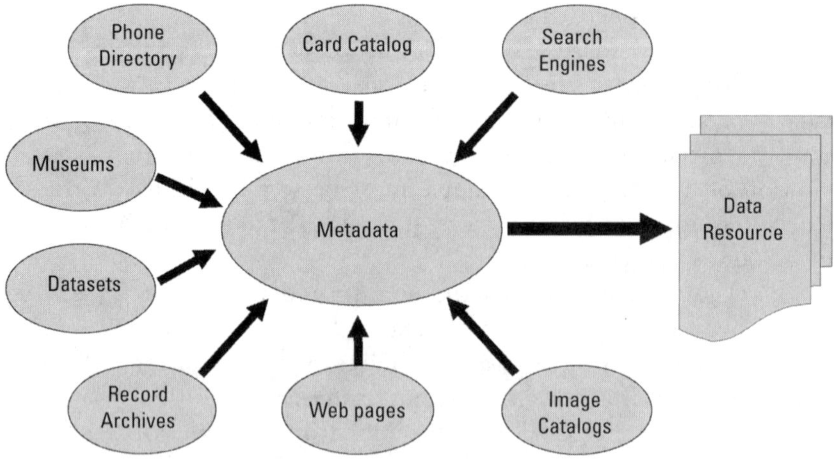

Figure 4.1 Various forms of metadata implementation and use.

metadata in the file headers. An archived map, on the other hand, may have a separate resource document containing the metadata for the map.

Embedded and separate metadata each have certain advantages. Embedding metadata within the resource ensures that the metadata will not be misplaced or lost and alleviates any problems with data-metadata linkage. Sometimes the capability to embed is not possible and, as a result, metadata has to be separate from the resource. This data separation offers easier resource location and retrieval, as well as offering a facile way to manage the metadata. Separate metadata are usually stored in a database and linked to the affiliate resource.

As you may have guessed, the topic of metadata is vast. In fact, it offers so many variations and covers such an array of fields that metadata can form (and has formed) the subject of its own book. Our interest in metadata goes only so far as the basics and those traits relevant to GIS.

Before jumping into the GIS-relevant side of metadata, it is important to understand the various objectives for the creation of metadata. The most primary objective for metadata is resource discovery. Metadata makes the resource "visible" to searchers, making it much easier for users to discover available resources. Other metadata goals include resource identification, enhanced structure, administrative controls, intellectual rights management, and preservation. These metadata objectives are equally important throughout the assorted fields of use.

Geospatial Metadata and GIS

We have already discussed how metadata are uniquely linked internally or externally to various types of data resources. In GIS, the data resource of paramount interest is the data set. Metadata that identifies and describes a data set, be it in GIS or another database management system, is classified as dataset metadata. Furthermore, since a GIS is a geospatial database utilizing a geospatial data set, its metadata are likewise geospatial metadata.

Users of geographic information systems rely heavily on the use of geospatial metadata for identifying and manipulating geographic data sets. Classified as dataset metadata, geospatial metadata are defined as structured information about a specific data set that enables the geospatial resource to be easily identified, used, and cataloged. Geospatial metadata includes at the very least the geographic data's owner, source, description, resolution, scale, and proper use.

Metadata helps GIS users understand the numerous parameters surrounding the data sets. Users can quickly discern the data set's level of precision and usefulness in tandem with other data sets and objects. Well-detailed geospatial metadata directs users on how best to use the resource and to what lengths the dataset creator went to construct the resource. Through metadata review, an astute GIS user could identify which dataset resources are of high quality, which are beneficial for their application, and which are pure garbage.

Geospatial metadata is used for the wide array of areas that GIS handles but has proved to be particularly valuable in the technical and scientific arenas. New, refined definitions of geospatial metadata are concurrently being constructed to better define and catalog the available technical resources. New pathways for information exchange and sharing have been paved through past metadata initiatives and continue to expand through ongoing initiatives within many progressive disciplines, such as in the biological, natural resource, and social science fields.

Every initiative to define and beneficially structure discipline-specific metadata is based upon an already established metadata format called a *schema* (or scheme). Refer to Figure 4.2 for an overview of a schema. Traditionally, *schema* is used for diagrammatic description and *scheme* is used for textual description. For the sake of simplicity and the nature of GIS data sets and products, we will use the term *schema.* Moving forward, the next section offers insight into metadata schemata and their element sets, as well as offering the particulars of the ones more universally employed.

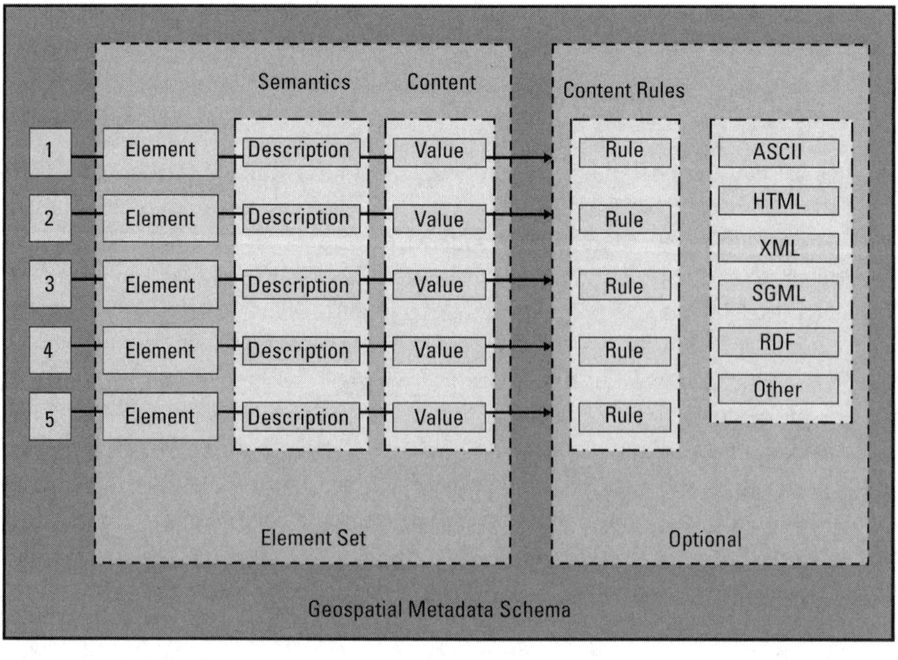

Figure 4.2 Geospatial metadata schema.

Metadata Schemata, Element Sets, and Syntax

A metadata schema is a description of the particular structure, arrangement, or content applied to a set of metadata elements. Alternately, the metadata element set is the collection of individual constituents that form the actual metadata. You can envision the metadata's schema as being the "database," the element set as being the database's primary table, and each element as being an individual data field within that primary table.

Metadata schema elements hold characteristics similar to any spoken or programmed language, whereby they must follow stringent rules for correct usage. The description of a schema element is considered the semantics of that schema and the values given to the schema elements are considered the content. This schema content may be further constrained through schema content rules. For instance, it may be required for a date-related schema element to be entered in an MM-DD-YYYY format. These elemental dictates are part of the overall structure of the schema and help maintain consistency throughout the developed resources.

Imposing further vigilance in metadata consistency, certain schemata must abide by well-defined content-encoding syntax rules. These rules offer a consistent metadata structure and further enhance the resource's discovery. As indicated, not all schemata are required to follow defined syntax rules and, as such, are called syntax independent.

To implement syntax, modern metadata schemata use various frameworks, such as ASCII, HTML, Standard Generalized Markup Language (SGML), Extensible Markup Language (XML), and Resource Description Framework (RDF). Figure 4.3 offers a few samples of metadata syntactic frameworks.

ASCII is the most basic of frameworks, simply because ASCII is nonsyntactic. ASCII is an original computer-based exchange format that is text only and is not capable of any embedded information. Therefore, however the metadata schema structures the data is how the data is displayed, without any content rules. ASCII proves to be minimally useful for resource discovery.

A better framework for geospatial metadata is one of the markup languages. HTML is a base framework that uses embedded element tags to describe the document. Although much more detailed than ASCII, HTML is rather simple and focuses solely on a document's style rather than its structure. XML, on the other hand, is an extension of HTML that uses tags to define and embed structure within the document. XML offers better versatility and enhances resource discovery and data interchange. In recent years, XML has become one of the syntactic frameworks of choice for metadata.

SGML departs from the role of essential resource description and structure to something more elaborate and comprehensive. SGML is an enhanced union of both HTML and XML and, as such, is often called an HTML-XML superset.

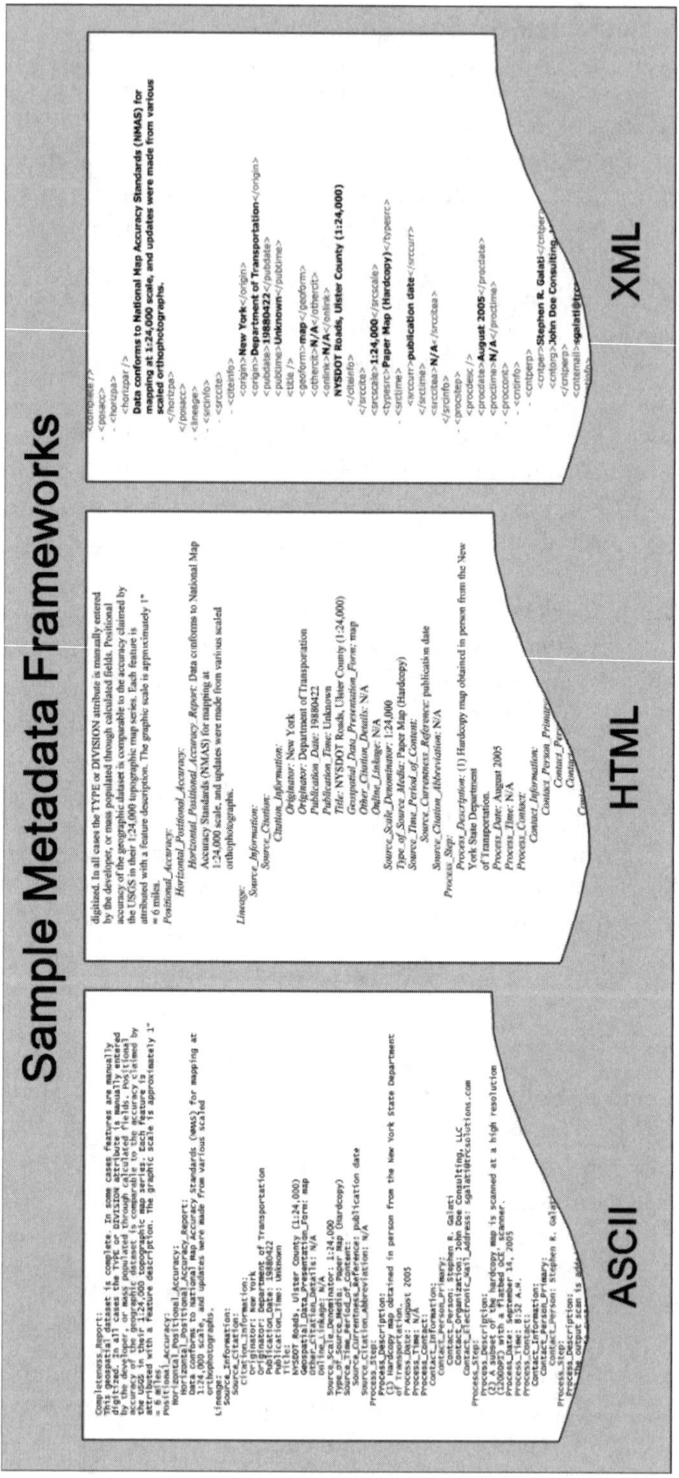

Figure 4.3 Sample metadata syntactic frameworks.

SGML looks much like an HTML document but is a structurally rich version of XML. With more structure descriptors, SGML offers copious description, both in style and structure, within the document. Additionally, the SGML framework offers the best availability for resource exposure and discovery among the discussed syntactic frameworks. However, with so much detail required, SGML is sometimes found too laborious. Figure 4.4 details the unique relationship between the markup language frameworks.

The final syntactic framework we will discuss is Resource Description Framework. RDF takes a different approach toward metadata architecture. As its name implies, RDF is a framework for describing a resource and sharing geospatial metadata. RDF is different from the markup language triad (HTML-XML-SGML) and focuses on the essential information that is needed to simplify metadata data searches, encourage interchangeability between syntaxes, and identify key resource properties.

In terms of depth of description, XML and RDF are the most similar but lack the comprehensive nature of SGML. You may be asking yourself, "Which one is better?" To be honest, it depends on the desired product, as well as the schema requirements, the desired amount of exposure, and the resource creator's effort to describe the resource.

For instance, some schemata (such as the Dublin Core discussed later) prefer RDF to one of the more traditional markup languages due to the ratio between metadata structure simplicity and essential content. Figure 4.5 depicts a sample RDF structure. Whatever syntactic framework and/or standard schema are used, the more description supplied will only benefit resource discovery, interoperability, and use.

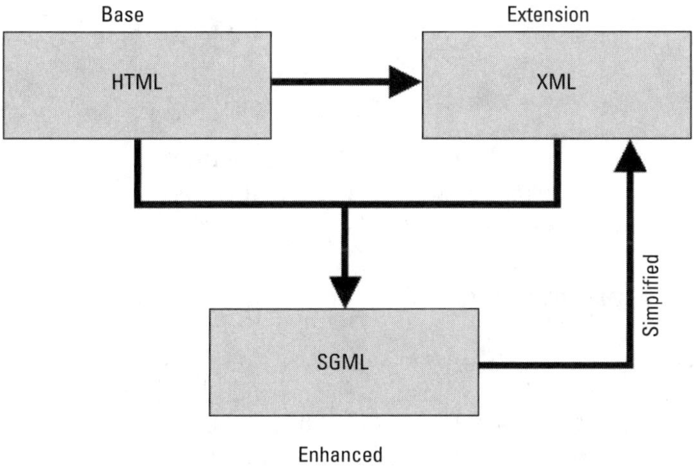

Figure 4.4 The relationship between HTML, XML, and SGML.

```
<?xml version="1.0"?>
<!DOCTYPE rdf:RDF SYSTEM "http://dublincore.org/documents/2002/07/31/dcmes-
xml/dcmes-xml-dtd.dtd">

<rdf:RDF
  xmlns:rdf="http://www.w3.org/1999/02/22-rdf-syntax-ns#"
  xmlns:dc="http://purl.org/dc/elements/1.1/">
  <rdf:Description rdf:about="\\metadata\standards\samples\rdfsample.html">

    <dc:title>
      GIS Demystified RDF Metadata Sample
    </dc:title>

    <dc:creator>
      Galati, Stephen R.
    </dc:creator>

    <dc:subject>
      Metadata.
    </dc:subject>

    <dc:description>
      Standards sample provides an example of Dublin Core standards
      as published in a Resource Description Framework form.
    </dc:description>

    <dc:publisher>
      Artech House
    </dc:publisher>

    <dc:type>
      Text
    </dc:type>

    <dc:format>
      text/html
    </dc:format>

    <dc:format>
      18425 bytes
    </dc:format>

  </rdf:Description>
</rdf:RDF>
```

Resource Description Framework (RDF)

Figure 4.5 A sample Resource Description Framework structure.

Geospatial Metadata Standards

Throughout the past years, metadata has become a hot topic for discussion. Many worldwide organizations feel that geospatial data should be standardized for swift interoperability and exchange, while others argue that many field- and location-specific structures are needed to keep metadata uniquely keyed toward the individual fields they represent. This also helps capture field-unique elements that a general standard would omit. As a result, there are numerous initiatives currently defining and instituting field-specific tangents (called profiles) off already established schemata.

Established and officially documented schemata form the current standards for today's geospatial metadata. Standards, as a whole, are valuable within the GIS industry for a number of good reasons. First, standards provide universal terms within the various resources they describe. This universal terminology offers beneficial consistency within the thousands (even millions) of available data resources within which the terminology is enforced.

Second, universal terminology presents the capability of automatic searches for specific terms. Users can program workstations to search for any universal term, such as *author, publisher,* and *source,* and save enormous amounts of discovery time. Automated searches can be done for nonstandard metadata, but these will take much longer and will habitually provide less than satisfactory results.

Third, similar to the second reason, users can quickly search numerous data sets and geospatial objects for specific terms. Having various forms of resources tagged with universal terms once again helps link the resources and produce comprehensive search results. Thus, finding specific information is usually much easier and time efficient within standardized metadata than within a nonstandard format.

Fourth, standards enable simpler information interchange. Since terms are set for individual metadata elements, the crossover between two standards is uniform throughout all associated resources. For instance, Dublin Core metadata can easily be transformed into European CEN/TC 287 metadata, since both are established element sets.

Fifth and finally, certain standards are mandated by governments and are required by local agencies. The U.S. government requires all agencies to use the Federal Geographic Data Committee content standard, while the Australian Government requires the Australian Government Locator Service standard.

Geospatial metadata standards offer the GIS user an efficient way to organize and catalog available images and data sets. As you may already grasp, there are numerous standards available to the worldwide public, with some schemata better for geospatial metadata than others. Some available standards for geospatial metadata include:

- Content Standard for Digital Geospatial Metadata (CSDGM);
- Dublin Core Metadata Element Set;
- European CEN Metadata Standard (CEN/TC 287);
- Australian Government Locator Service (AGLS) Metadata Element Set;
- UK GEMINI Discovery Metadata Standard.

To get a better idea of metadata structure and how each schema differs, we take a brief look at these metadata keystones.

Content Standards for Digital Geospatial Metadata

Under an executive order from the U.S. government, the Federal Geographic Data Committee (FGDC), a 19-member interagency committee, was charged with the task of developing national geospatial metadata standards. As a direct result, the FGDC developed the Content Standards for Digital Geospatial Metadata, or CSDGM. Since its origin in the mid-1990s, the CSDGM has developed into the most widely used metadata standard to date. CSDGM has even become the basis for numerous standards worldwide.

The intent for CSDGM was to help the U.S. government minimize the costs of data acquisition efforts and to promote interagency data sharing. The CSDGM was developed for consistent geospatial dataset description and utilizes a group of seven core (primary) and three floating elements. The core elements make up the major element set, whereas the floating elements are nonessential and wholly optional. Figure 4.6 details the seven CSDGM metadata core elements.

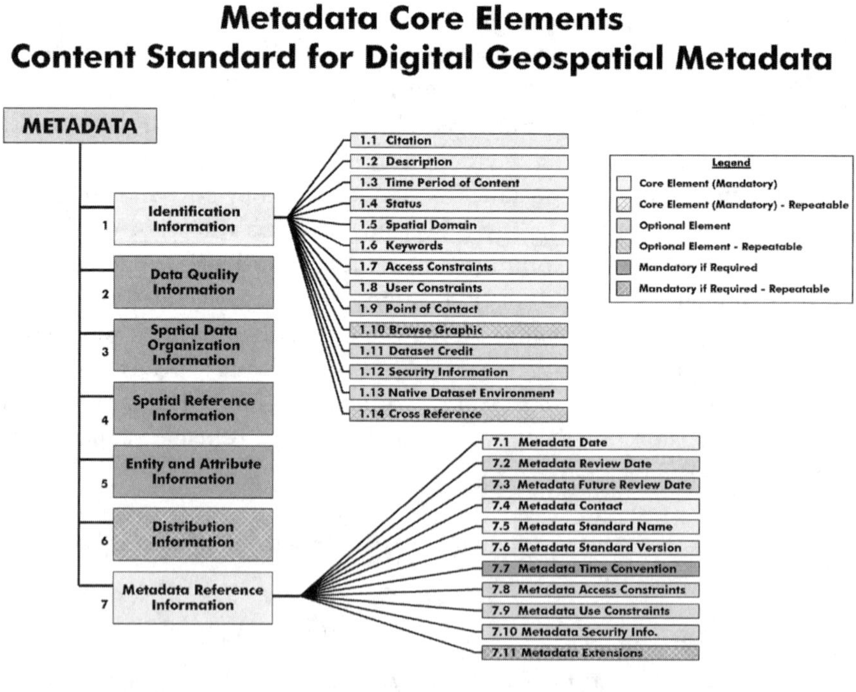

Figure 4.6 CSDGM metadata core elements.

The core elements are separated as *Mandatory, Optional,* and *Mandatory if Required.* The *Mandatory* elements are required in order to be minimally compliant with FGDC's standard. The *Optional* elements can be added or omitted from the metadata without affecting metadata compliance. The *Mandatory if Required* elements are metadata elements that may or may not be applicable, depending upon the resource being described. For instance, CSDGM metadata for a Web page may warrant only the minimal core elements, while a Geocoded map may also require one or more *Mandatory if Required* elements, such as data quality information. Although Figure 4.6 breaks down only the two *Mandatory* elements into their subcomponents, each element has a similar breadth of subsets.

The *Mandatory if Required* status of CSDGM elements adds a level of user flexibility and decision making. As with any form of metadata, the more description included in metadata, the greater the discovery potential. Hence, if the resource information is available to the metadata creator, it is in his or her best interest to include that additional description.

Certain elements and element subcomponents are repeatable, which means that they can appear in metadata more than once. Say, for instance, there was a set distribution for a particular metadata (e.g., 14 college professors). You have the capability within the CSDGM structure to include information about each of the 14 professors within the *Distribution Information* element. As depicted in Figure 4.6, this core element is repeatable and, thus, would be compliant with FGDC's standards.

As a major, widely available metadata element set, CSDGM sets the bar for geospatial metadata consistency and standards and continues to encourage international data exchange. As the leader in the geospatial metadata arena, its structure has been emulated for countless standardized element sets around the world. Adding credence to CSDGM is the fact that it still remains the national standard for the United States.

Dublin Core Metadata Element Set

The Dublin Core Metadata Element Set began as an initiative to advance the discovery of information resources. The initiative, called the Dublin Core Metadata Initiative, started in 1995 with a workshop in Dublin, Ohio, and involved various types of interested professionals, such as researchers, librarians, and content providers. The result of this initiative was the creation of a Dublin Core standards prototype, which involved a rather small group of information descriptors. The Dublin Core standards quickly gained worldwide attention and interest from a wide variety of metadata users.

Today, the Dublin Core Initiative's ongoing progress has produced the Dublin Core Metadata Element Set, a 15-element set that has attained worldwide acceptance and use within a wide range of content areas. The Dublin Core sets the standard for multidomain resource description and data resource visibility.

Figure 4.7 represent the 15 primary elements that make up Dublin Core metadata. Unlike the CSDGM element set, all Dublin Core elements are optional and repeatable and may be used in any order. Therefore, the primary elements depicted in Figure 4.7 are in no particular sequence and can be changed, duplicated, or omitted from metadata. This innate flexibility has made the Dublin Core standards highly attractive to users worldwide, as well as an excellent schema for geospatial metadata.

Going way beyond the Dublin Core Initiative's original expectations, the Dublin Core Metadata Element Set is a means for individual preference and metadata richness. The element set offers the use of qualifiers, which are bits of additional information that are useful in further exposing the resource for discovery and promoting enhanced information interchange. Qualifiers include such additional information as various encoding syntaxes (i.e., HTML, XML, and RDF), greater content descriptors, and value type description. In fact, Dublin Core users can choose between Simple Dublin Core metadata and Qualified Dublin Core metadata.

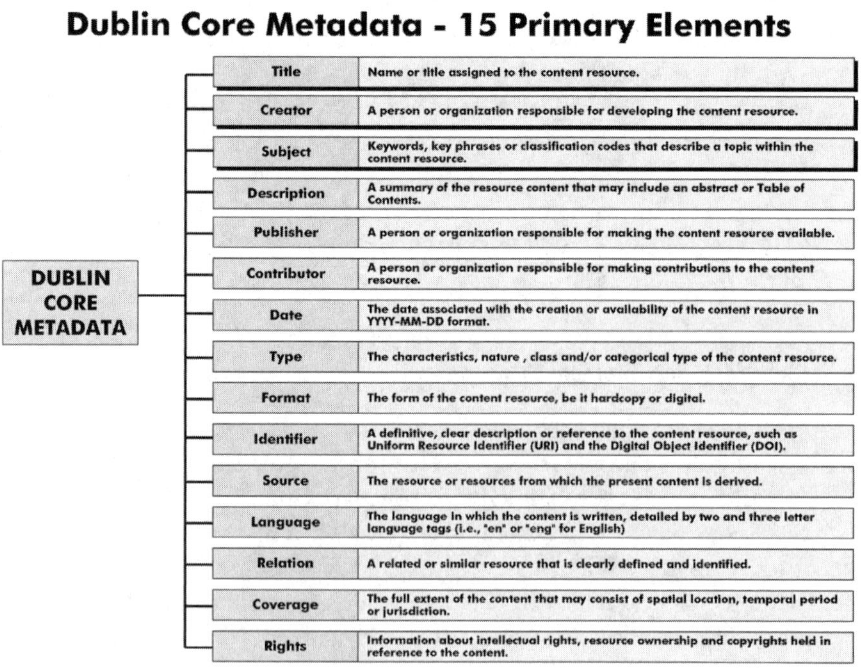

Figure 4.7 Dublin Core metadata—the 15 primary elements.

Simple Dublin Core metadata, also known as Unqualified Dublin Core metadata, involves the use of only the 15 primary elements with simple content and/or values. Qualified Dublin Core metadata goes beyond the simple content within the 15 primary elements and enables the metadata writer to use qualifiers to add a greater level of resource detail. Both forms of Dublin Core standards are efficient, but in some instances Qualified Dublin Core can reduce the interoperability of the metadata and add an unwanted layer of complexity.

As mentioned earlier, the Dublin Core Metadata Element Set also allows for various encoding syntax, which helps interchange Dublin Core descriptors into other metadata schemata. When Dublin Core standards first came out, they were predominately encoded in HTML. Nowadays, the prevailing encoding is XML and RDF, whereby XML is typically used for Simple Dublin Core metadata and RDF for the more complex products. Whatever format Dublin Core metadata takes, its beneficial flexibility and cost-effective simplicity has made it an internationally recognized metadata leader.

European CEN Metadata Standard

Departing from the individual "national standard" ideology, most European countries follow the European CEN Metadata Standard known as the CEN/TC 287 metadata standard for geographic information. Since the late 1990s, CEN/TC 287 has become the most widely adapted geospatial model in Europe that is fully maintained and standardized through a technical committee.

The CEN/TC 287 metadata standard is governed by the Comité Européen de Normalisation (European Committee for Standardization), or CEN for short. CEN was founded in 1961 and contributes to the standardization, exploration, and development of various research programs. It utilizes special technical committees (TCs) for each of its program areas, with TC 287 dedicated to geographic information.

Differing from the FGDC's and Dublin Core's hard-nosed structure, the CEN model offers flexibility with the number of different elements. For instance, the Dutch version of CEN/TC 287 (called NVN-ENV 12657) contains nearly 300 metadata elements. Interesting enough, if you were to compare, side by side, CEN/TC 287 to either the FGDC or Dublin Core standards, you could easily link one CEN/TC 287 element tag to many FGDC/Dublin Core element tags.

Now, GIS catalog software producers are taking notice of the European CEN Metadata Standard. ESRI, in particular, has developed a way to link CEN/TC 287 standards and apply them to FGDC's format. Their ArcCatalog software, for instance, takes CEN metadata and transforms it into the CSDGM schema in XML. As more and more software producers implement the CEN model, these standards will gain even more popularity within Europe and, possibly, elsewhere.

In recent years, the initial publishing of ISO/TC 211, a widely accepted international geospatial metadata standard, brought the work of Technical Committee 287 to halt. As the ISO/TC 211 metadata standards were disseminated and implemented, CEN realized that TC 287 was again needed, but this time to help implement the new ISO standards within Europe through the CEN model. Therefore, the European CEN Metadata Standard is still one of the industry's major metadata leaders. For more information about the CEN standard, visit their Web site at http://www.cenorm.be.

Australian Government Locator Service Metadata Element Set

The Australian Government Locator Service (AGLS) Metadata Element Set first developed from a National Archives of Australia workshop in late 1997. Since then, the AGLS Metadata Element Set has become the leading geospatial metadata standard used throughout Australia. The AGLS element set is designed to enhance the exposure, discovery, access, and interchange of available government resource information and services.

The AGLS element set is a group of 19 descriptive elements that are intrinsically based upon the 15 element descriptors from the Dublin Core Metadata Element Set. The AGLS element set, however, is much more complex, since it allows richer element description through element qualifiers. These qualifiers enable the elements to describe a greater array of resource categories.

Because the AGLS element set is a derivative of the Dublin Core element set, both are entirely compatible with each other. This fluid data interchange adds to the greater endeavor of universal metadata standardization and enhanced resource access. The AGLS Metadata Element Set remains a modern and attractive standard that not only operates seamlessly with other standards but also adds a discriminating level of geospatial resource description.

The UK GEMINI Discovery Metadata Standard

The UK GEMINI Discovery Metadata Standard is a defined metadata element set used within the United Kingdom for describing geospatial data for faster, more accurate searches (metadata discovery). As such, this U.K. standard focuses entirely on discovery-level metadata, which is the specific metadata responsible for expediting resource information retrieval. The UK GEMINI standard is an element set derived from ISO 19115 (Geographic Information—Metadata) and the U.K. e-Government Metadata Standard (e-GMS). This metadata standard has gained industry respect and use, since it is nourished by the government, academia, and the private sector geospatial community.

The UK GEMINI Discovery Metadata Standard is a distant relative of the Dublin Core standard, since it has been developed to match explicitly with the element tags of e-GMS, a standard based upon the Dublin Core. The UK GEMINI standard uses 32 elements, ranging from *Title* and *Data Format* to *Use Constraints* and *Lineage*. Similar to the CSDGM, there are mandatory and optional elements, specifically 17 mandatory and 15 optional elements. Some of these elements have just a single occurrence, while others are repeatable.

Adding to the level of discovery the UK GEMINI standard offers is the depth of description and information attached to each element. Each element not only has the typical name, description, and value but also has extra description information including equivalent e-GMS and ISO 19115 element names and the element's data type. Given the added information, the UK GEMINI standard fulfils its main purpose of heightened geospatial metadata discovery.

Executive Orders 12906 and 13286

In the 1990s, the U.S. government became very interested in the retention, organization, and dissemination of geospatial data. By this time, GIS, GPS, and satellite imagery were being used in full force throughout the federal and local levels of government. As public interest in natural resource stewardship, environmental protection, and economic development soared, so did the plethora of data resources, duplication of data efforts, and inconsistencies in resource retention. It became clear to the government that standardization and assemblage of data resource dissemination was desperately needed.

Acting as a geospatial impetus, Federal Circular No. A-16, originally published in 1953 for survey and mapping issues and revised in 1967, was again revised in 1990 to include spatial data. This second revision marked the origin of the FGDC and took the first steps toward a comprehensive National Spatial Data Infrastructure (NSDI).

On April 13, 1994, President Bill Clinton signed Executive Order 12906, which was a giant step toward coordinating geospatial data acquisition and access. In it, the NSDI was formed and a mandate for the FGDC to create geospatial data standards was enacted. FGDC's Content Standard for Digital Geospatial Metadata was essentially born.

Even more specific to GIS, Executive Order 12906 detailed plans for a national digital geospatial data framework, whereby all federal agencies and organizations were required to document their geospatial data using the FGDC's standard. Additionally, all published metadata now had to be distributed and made available through a clearinghouse system. As you can imagine, this executive order proved to be the push needed to set the geospatial community in motion.

Contributing to the release and dissemination of geographic and geospatial information came a second significant executive order from the U.S. government. On March 5, 2003, President George W. Bush signed Executive Order 13286, which amended Executive Order 12906 to represent changes in government, as well as to represent changes in geospatial management, changes in technology, and clarifications to FGDC responsibilities. Without question, these federally imposed geospatial dictates helped to build the enormous data repository available today and to disseminate the importance and efficacy of geospatial metadata.

The National Spatial Data Infrastructure

With the signing of Executive Order 12906, the FGDC was charged with the task of developing the NSDI. The original concept was to have a centralized architecture for geospatial data policies, standards, and procedures. The NSDI was the conceived solution to disseminate geospatial data in a consistent fashion, while flattening the growing costs of duplicated governmental geospatial data collection.

As part of the vision, NSDI would provide an infrastructure through which data producers and users could share geospatial data. This producer-user relationship has already started to emerge from FGDC's efforts. Through the NSDI, the FGDC has begun to make key partnerships with public and private data producers to increase data availability to geospatial data users.

Incidentally, one valuable outcome of the NSDI is that it affords emergency services the capability of accessing geospatial data in an easy, expedited manner. In emergency situations, such as natural disasters, terrorism, and accidents, expedited and consistent data retrieval can truly hold life and death in balance. As the NSDI continues to develop and grow, the scope of available benefits and innate advantages will become ever clearer.

Although in recent years the FGDC has made great strides in implementing the NSDI, more work is needed. NSDI is still a work in progress. FGDC's effort is ongoing and is in cooperation with many organizations from state, local, and tribal government levels as well as the academic community and the private sector. The U.S. government places great trust in the fact that this infrastructure will provide substantial cost savings and facilitate enhanced decision-making. The truth of the matter is that the NSDI is still in its early stages and, in the coming years, may prove to be the resource that was originally envisioned. As with all things, only time will tell.

Geospatial Metadata Clearinghouses and Publishing Resources

As mentioned previously in this chapter, the U.S. government mandated that all published geospatial metadata be made available through a clearinghouse system. A metadata clearinghouse is an online or offline reservoir of published metadata that is being offered to the public. These clearinghouses in general offer freely distributed metadata, although some provide for-purchase variations of hard-to-obtain information.

Table 4.1 details just a small selection of international geospatial metadata clearinghouses. All of these sites have free metadata. There are, however, hundreds of available metadata clearinghouses available, mostly online.

Often these geospatial metadata clearinghouses also offer free and for-purchase geospatial data sets. We discuss these particular geospatial data clearinghouses in more detail in Chapter 15. For now, focus on the geospatial metadata these data bureaus provide and understand that the best way to create your own metadata is to examine what has been done by industry peers.

Take some time to explore the clearinghouses detailed in Table 4.1. You can gain a great deal of understanding by examining the language, syntax, and structure of a few good metadata samples. In fact, these samples will serve as distinct resources as to how compliant metadata are written.

If you want to try your hand at writing metadata, a good place to start is with the U.S. Geological Survey and, in particular, Peter Schweitzer's work with metadata. Mr. Schweitzer has written numerous documents on metadata basics and created software to help people write formal metadata (Unix-based XTME and Windows-based TKME editors), preparse it for proper structure [i.e., "chew and spit" (CNS) metadata preparser], and parse it for compliance with FGDC standards [metadata parser (MP) software].

In addition to the FGDC-compliant software, each metadata standard (e.g., Dublin Core) has its own editors and parsers for compliance with its standards. Check out the particular standard's Web site for more information on its available software and requirements. Once you become more comfortable with metadata and GIS, you will be able to create your own metadata and publish it to a clearinghouse.

Table 4.1
Geospatial Metadata Clearinghouses and Publishing Resources

Nairobi Metadata Explorer (GRID)	http://gridnairobi.unep.org/metadataexplorer/explorer.jsp
Colorado Plateau Environmental Metadata Clearinghouse	http://mprlsrvr1.bio.nau.edu/metadataexplorer
Eastern Sierra Geospatial Data Clearinghouse Metadata	http://www.wmrs.edu/resources/data%20access/clearinghouse/contents.htm
EPA Geospatial Data Clearinghouse	http://www.epa.gov/nsdi
FGDC National Data Geospatial Clearinghouse	http://clearinghouse1.fgdc.gov
Geospatial One-Stop Portal	http://www.geodata.gov
National Biological Information Infrastructure (NBII) Metadata Clearinghouse	http://www.nbii.gov/datainfo/metadata/clearinghouse
University of Sydney's ECAI Clearinghouse	http://ecaimaps.berkeley.edu/clearinghouse/
NSDI Geospatial One-Stop metadata publishing guidance	http://www.geo-one-stop.gov/metadata
National Oceanographic and Atmospheric Administration (NOAA) Coastal Data Locator	http://www.csc.noaa.gov/cgi-bin/id/cid2k/cid2k.cgi?page='cdl'
Tutorials on setting up a clearinghouse node	http://www.fgdc.gov/clearinghouse/tutorials/howto.html
U.S. Geological Survey (USGS) XTME metadata editor (Unix)	http://geology.usgs.gov/tools/metadata/tools/doc/xtme.html
USGS TKME metadata editor (Windows)	http://geology.usgs.gov/tools/metadata/tools/doc/tkme.html
USGS CNS metadata preparser	http://geology.usgs.gov/tools/metadata/tools/doc/cns.html
USGS MP metadata parser	http://geology.usgs.gov/tools/metadata/tools/doc/mp.html

Part II: Geodesy, Earth Models, and Coordinate Systems

Understanding the Core GIS Environment

5

The Basics of Geodesy and Scale

The story of Christopher Columbus and his bold testament to the Spanish monarchy that the world was round and not flat is captivating and has remained an interesting tale within world history. Learning of Columbus as a child, I remember thinking: how could they believe the Earth was flat? I now look back and realize that science had proved the common understanding of the Earth's shape long before you and I were born. In fact, we were born into a society holding common and mature geographic knowledge, much different from Columbus, who was born into a society with varying images of the Earth, abundant misunderstandings, and science in its infancy.

So where does that leave us in terms of GIS? Well, for starters, through the discussions in Part I, we have begun clarifying the role of GIS by way of a primary understanding of GIS's groundwork. Let us now go further, tackling the core of GIS, the nature of geographic data, and the backbone to correct geographic representation and structure. Therefore, the science behind the Earth is our next destination, since it serves as a logical path to understanding GIS and its employment within our society. For this, we are directed toward the study of the Earth and the principles behind its shape, its size, and the laws that govern its portrayal. Let us begin Part II with geodesy.

What Is Geodesy?

Geodesy is the division of science associated with the measurement and portrayal of the Earth. Geodesy, known also as *geodetics,* is intimately concerned with the determination of the size and shape of the Earth, as well as its elements.

These Earth-based elements include its terrestrial gravity, magnetic field, tides, geologic and crustal movement, and polar motion.

As a study of the Earth and its rudiments, geodesy covers a large analytical landscape and combines the sciences, mathematics, physics, and observation. This analytical landscape and questions surrounding Earth's geodetic elements first initiated the science of geodesy. The study of geodesy began as mere curiosity and an undying human will to explain the Earth's unknowns through logic. There exist the first remnants of geodetic analysis starting as far back as the early Greeks. Homer, Pythagoras, Plato, and others all had ideas about the shape of the Earth. Homer, for instance, held the idea that Earth was a large flat disc, while Pythagoras, a mathematician, viewed Earth as an idyllic spherical figure. Plato logically guessed that the circumference of the Earth was 40,000 miles.

This early form of geodesy was based upon visual observations, logical assumptions, and crude calculations and transformed the notion of a flat Earth to the idea of a curved surface. In fact, Greek sailors took notice that as they came closer to land the visible high points seemed to ascend from the water. Others noticed that astronomical events such as a lunar eclipse occurred high in the night sky on one end of the Mediterranean while on the other end of the Mediterranean it occurred closer to the horizon. These observations added reason to the common consensus in a round or curved Earth.

Many, like Plato, made educated guesses as to the size of the Earth, but they were just guesses. Archimedes went one step further, providing an educated approximation that the circumference of the Earth was 30,000 miles. However resolute, no one could definitively prove the Earth was round or determine its true size. Along came Greek mathematician Eratosthenes who with curious mind and mathematical craft created a method to answer these geodetic unknowns.

Eratosthenes had taken notice during a summer solstice (longest day of the year) that a stick placed straight up in the sand cast no shadow and that the midday sun (directly overhead) shone to the very bottom of a well. Both observations occurred in his home town of Syene (now Aswan), Egypt. Eratosthenes also noticed that at the same time, the sun was not directly overhead in Alexandria, a town roughly 500 miles away. As a result, a shadow was cast against the vertical in Alexandria that was equal to 1/50th of a circle or 7 degrees 12 minutes ($7^\circ 12'$). See Figure 5.1.

Eratosthenes took his summer solstice observations and applied some assumptions: (1) The distance between Syene and Alexandria was 500 miles; (2) since Syene's midday summer solstice sun was directly overhead and, on that day, the sun moves along the tropic of Cancer, it was concluded that Syene was on this tropic zone line; and (3) Syene and Alexandria lie on a direct north-south line.

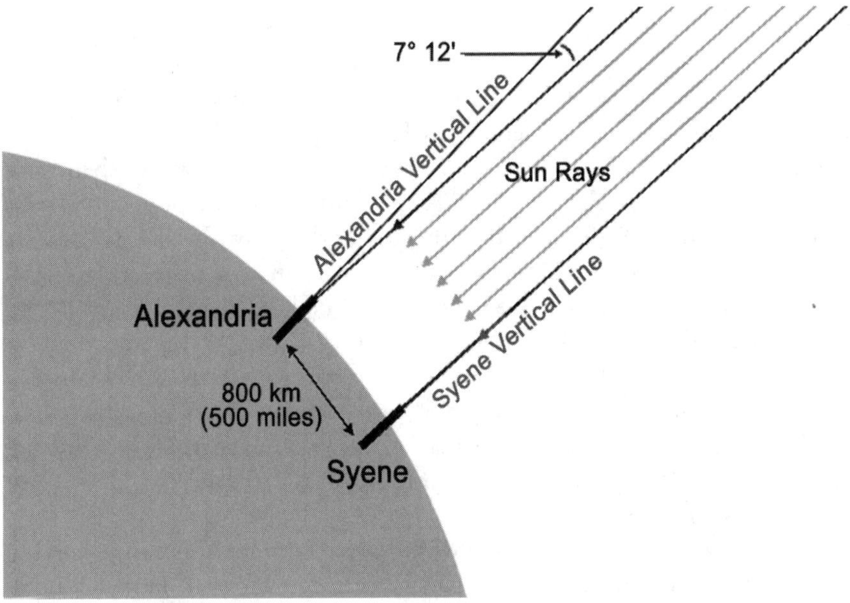

Figure 5.1 Eratosthenes's method of calculating the Earth's size.

Based upon these assumptions and the measured shadow, Eratosthenes concluded that the distance between Syene and Alexandria was 1/50th the circumference of the Earth. Mathematically it can be explained as:

$$1/50 \times \text{Earth's circumference} = 500 \text{ miles}$$

$$\text{Earth's circumference} = 50 \times 500 \text{ miles} = 25{,}000 \text{ miles}$$

Remarkably, Eratosthenes made a nearly accurate measurement of the Earth's size given that today's accepted World Geodetic System (WGS) calculation is 24,901 miles! However, Eratosthenes's calculations were somewhat incorrect, since every assumption was a bit erroneous. We now know that (1) the real distance between Syene and Alexandria is 453 miles; (2) Syene lies 37 miles north of the tropic of Cancer line; and (3) Syene lies 3 degrees 30 minutes (3° 30′) east of Alexandria's north-south line. Additionally, Eratosthenes's calculation of 7° 12′ from the vertical is actually 7° 5′. Despite these facts known in hindsight, Eratosthenes became the first to substantiate the calculation of Earth's size.

After Eratosthenes, others continued to revise the Earth's measurement with varying results. It was not until after the invention of the telescope, development of the logarithm tables, and progressive understanding of the Earth's true shape that more exact Earth size measurements were calculated. With these new measurements and Earth size-shape concepts came the maturation of geodesy and the development of the current conventions.

Present-day geodesy is well beyond questioning Earth's overall shape and size; it now focuses upon Earth's actual shape with its unique physical terrain. Geodesy centers upon Earth's significantly vast landscape with its drastic elevations and varying land forms. This deviating terrain is a true surface and includes high mountain ranges, sweeping plains, and variable geologic surfaces. This true surface adds countless levels of scientific complexity to geodesy and, in conjunction with the desire to measure and represent the Earth, forces concern about both accurate Earth positioning and the determination of Earth's actual shape. Thus, reducing the Earth to a simple shape (i.e., a spheroid) is of menial scientific interest in current-day geodesy other than for simple calculations and geographic representation.

Present geodetic concerns subsist within a three-dimensional time-varying space. This means that over time the Earth's surface changes due to temporal surface variations, geologic events, the gravitational field, and other surface-altering episodes. Erosion, for instance, will permanently change a coastline over time, or a river basin swollen by excess storm water runoff will temporarily alter features on the Earth. Both, however, alter the true shape of the Earth at any given time. These sometimes unpredictable changes add layers of geodetic complexity to the accurate measurement and representation of the Earth.

Geodesy has also advanced and now, to some extent, relies upon two forms of terrestrial measurement: geomensuration and surveying. *Geomensuration* is the measurement of the Earth as a whole (i.e., on a global level), while *surveying* is the measurement of individual parts on the Earth's surface (the ever-changing physical terrain). Both techniques offer credible measurements wholly useable toward accurate geodetic positioning, surface shape, and size. GIS employs these forms of terrestrial measurement through various data sources, such as satellite, GPS, and field measurements. As such, geodesy serves as an important science and elemental backbone to GIS.

With this understanding, we are now ready to discuss the pertinent geodetic elements that define geodesy and provide crucial pieces of geographic information. Specifically, the reference Earth shape (the ellipsoid), geodetic positioning (geodetic datum and coordinates), the true Earth shape (the geoid and vertical datum), and the usable portrayal of the Earth (map projections). The next four chapters go into great detail about these geodetic elements; however, before we move into these areas, we must first discuss the principles of map scale. In short, map scale provides the exact geodetic relationship between the GIS information product and reality.

Understanding Map Scale

You are driving on a desolate back road in Montana and your car's gas light turns on. Nervously you reach down and grab a road map. You easily find your position and pan the paper map for the nearest town that would have a gas station. Three inches from your position is a reasonable sized town by the looks of it. From experience you know that when the gas light turns on, there is enough fuel in the tank for another 40 miles. Three inches isn't that far, right?

As the dread of waiting for a tow truck fills your mind, you notice in the corner of the map the words, "One inch represents 20 miles." You swiftly conclude that 3 inches must represent 60 miles and that you can reach the town before running out of gas. You continue toward town with newfound peace of mind.

We have all been in a similar situation, though maybe not as dramatic. The group of words that ultimately informed and relieved the Montana traveler is called a map scale. A map scale, in its most common treatment, relates to the size of map-based features in relation to their actual, real-world size. Specifically, map scale is the relationship between a distance on a map and its corresponding distance on the ground.

Although detailing the same information, not all map scales look alike. In fact, there are three common forms of map scale: verbal scale, graphic scale, and ratio scale. Refer to Figure 5.2 for a graphic illustration of each discussed map scale.

Representative Fraction	Verbal Scale
1:5,000	One inch represents 1 mile
1:25,000	One inch represents 3,000 feet
1:50,000	One cm equals 40 kilometers
1:250,000	One quarter inch equals 400 inches
1:1,000,000	

Graphic or Bar Scale

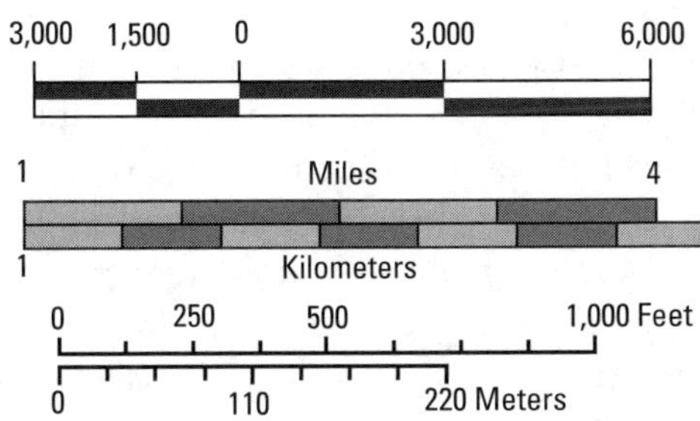

Figure 5.2 Key map scales.

In the example Montana situation, a *verbal* scale was used. Verbal scales express the map-to-ground relationship in words, such as "One inch represents 20 miles" or "One inch equals 20 miles."

A *graphic* scale or *bar* scale depicts a bar of a certain length that represents a distance length. Graphic scales enable the map user to visually recognize distances with relative ease. Because of this simple understanding, graphic scales are undeniably the most used form of map scale. One striking feature of the graphic scale over verbal and ratio scales is that graphic scales remain accurate if the map is photographically enlarged or reduced.

A *ratio* scale, more formally known as representative fraction and less formally as natural scale, reduces a map's distance to 1. One unit on the map corresponds to a distance in the same units on the ground. In a typical 30-minute series topographic map, the representative fraction is shown as either "1/125,000" or "1:125,000" whereby the 1 and 125,000 are the same unit of measure.

When dealing with GIS products, the term *scale* can also be used in a wholly different manner. The term holds significance to describe the working size of an application or project area, specifically *small scale, mid-scale,* and *large scale.* Parameters for each distinction are highly flexible, based upon individual judgment rather than a specific criterion.

Small-scale applications involve relatively large areas, such as a map of the world, the continent of Australia, or the city of Rome, Italy. A typical small-scale representative fraction can be assumed anything greater than 1:500,000. Large-scale applications involve small areas of focus, such as a community, a city block, or a university campus. A typical large-scale representative fraction can be assumed anything less than 1:50,000. Mid-scale applications, on the other hand, fall right in the middle of large- and small-scale applications, with a typical representative fraction ranging from 1:50,000 to 1:500,000.

As stated earlier, parameters for this scale nomenclature are determined by the individual user or the specific application. What is considered large scale for one application may be considered as mid or small scale on another. Figure 5.3 presents an example of this flexible scale nomenclature. Notice that the relationship between representative fraction and scale nomenclature is different from our previously stated "typical" ranges. This specific application's overall area covers roughly 20 square miles with the small-scale view covering the entire area and the large-scale view covering less than one square mile.

It is easy to get confused with this scale nomenclature, thinking a small-scale application should show only a small area. This, of course, is wrong. There is a rule of thumb to remember when using this scale terminology: Large scale has a large representative fraction [i.e., the denominator (number after the colon) is small] and small scale has a small representative fraction (i.e., the denominator is large).

1 : 126,000
Small Scale

1 : 61,400
Mid-Scale

1 : 7,880
Large Scale

Figure 5.3 Large-, mid-, and small-scale applications.

Simple Map Scale Calculations

Regardless of specific format, map scale can be calculated using the following equation:

Map scale = Distance on map/Distance on the ground

In the Montana example, every map inch (distance on the map) represented 20 actual miles (distance on the ground), thus, map scale = 1/1,267,200.

There are many instances when map scales need to be converted into another form, such as a verbal scale to a representative fraction or vice versa. These scale transformation calculations are simple and prove worthwhile for various cartographic and GIS applications. Let us take a look at these calculations in Table 5.1.

Scale and Accuracy

An important characteristic of geographic data is that it can be represented differently at different scales. Proper map scale within GIS and on hardcopy maps enables the user to display accurate geographic data that is easily discernable, representative of the particular application and wholly successful in its implementation. Geographic features that are prominent at 1:24,000 scale are many times indistinguishable at 1:1,000,000. As with any GIS function, a user's knowledge of the final use is pertinent to the quality production of geographic information products.

Table 5.1
Scale Transformation Calculations

Conversion from Verbal Scale to Representative Fraction
1 inch represents 4 miles
1 inch equals 4 miles
1 inch = 4 miles
1 inch = 4 miles × 5,280 feet/mile
1 inch = 21,120 feet × 12 inches/foot
1 inch = 253,440 inches
1:253,440
Conversion from Representative Fraction to Verbal Scale
1:100,000
1 inch = 100,000 inches
1 inch = 100,000 inches × (1 foot/12 inches)
1 inch = 8,333.33 feet × (1 mile/5,280 feet)
1 inch = 1.578 miles
1 inch represents 1.578 miles

For specific applications or to develop certain targeted information products, geographic data representation can be manipulated by increasing or decreasing map scale. For instance, state planners targeting an individual land parcel would employ parcel maps at 1:24,000 or larger scales. If they were targeting regional land parcels instead, area maps at 1:250,000 or smaller scales would be the most useful.

When scale is adjusted by enlarging or minimizing the area view, accuracy of the data becomes a concern. Typically, as scale decreases, the number of specific features and the detail of those features also decrease. As such, data accuracy is lessened to the extent that the overall representation is more general than descriptive. When scale of the original data is increased, accuracy is not improved, but features are enlarged.

For example, take another look at Figure 5.3. The frozen pond prominent in the large-scale view (1:7,880) is but an indistinct feature among the landscape when scale is decreased to a small-scale view (1:126,000). As apparent, the amount of feature (pond) detail in the large-scale view is much greater than in the other two views. Consequently, when the scale gets reduced (large to small), feature detail becomes too little to see and cannot be adequately represented.

Proper map scale is directly proportional to accuracy. Sometimes when map scale is decreased and maps become small scale, geographic detail is lost. This can cause serious user problems, since geographic unknowns arise and accuracy diminishes. For instance, imagine if a person new to Interstate 81 in Virginia used a map of North America instead of a state map to get from Pulaski to Harrisonburg. The user may or may not get to his destination, since the North American map is dramatically less accurate in terms of Interstate 81.

In short, scale and accuracy go hand in hand with both GIS-based products and hardcopy cartographic maps. Every data source has accuracy limitations, and users must make every attempt possible to define the legitimacy of data sources. If scale and accuracy are used appropriately, the geographic information product will be used as intended and offer a well-defined and highly efficient representation.

Geographic Data Generalization

Geographic technologies represent the real world mainly for convenience, similar to the role played by hardcopy paper maps used by our wandering ancestors. For example, a traveler cannot simply fold up a true representation of his 100-mile route and stick it in his coat pocket. For the sake of convenience and portability, the 100-mile route is reduced to a scale of 1:100,000 (i.e., 1 inch equals 100,000 inches), which is then drawn and portrayed on paper, folded, and placed in his pocket. This diminution of scale is adequately achieved through generalization.

Generalization at its core is a method to manage and reduce unwieldy or overly numerous input data sets. Through this process of reduction, the real-world data source is rendered in less detail and, as a result, is somewhat distorted from reality. Each level of generalization introduces a greater degree of distortion; however, immediately, generalization affects the overall appearance and, even more important, the precision of the output data. Smart geographic data users are aware of the precision of data they use because it has a direct correlation to and effect upon the ultimate usability of their geographic information products.

Even by today's technological standards, an extremely hefty 500-gigabyte high-resolution image is not a manageable quantity of input data. Like trying to eat a 6-foot sandwich all at once, it is simply not feasible. So what is considered a manageable quantity? Every user has an individual parameter for manageability, which takes into consideration the computer and hardware specifications, the patience for waiting out long processing tasks, and the ultimate goal of analysis. A median case could potentially involve generalizing the 500-gigabyte image to a more tolerable 1.2-gigabyte image, resulting in a sleeker file more adaptable and friendly to geoprocessing.

All geographic data are in some way generalized. This is an inevitable, astringent process that alters the actual character of the data into a simplified and shrunken model of the real world. Ultimately, scale dictates the severity of generalization, varying in level from project to project. For instance, a map drawn to 1:100,000 scale (smaller scale) is more generalized than the same area drawn to 1:24,000 (larger scale), thus the smaller the scale, the greater the level of generalization. Chapter 11 take a much closer look at scale factors and the resultant degrees of distortion.

Figure 5.4 exemplifies geographic data generalization, wherein a physical structure is portrayed in relation with two different interpretations of the same roadway.

Figure 5.4 illustrates the quintessential "house on the wrong side of the road" problem. The 1:50,000 scale details a generalized road condition and offers severely less detail. The 1:10,000 scale details a close representation of the actual roadway, with the appearance of blatant roadway shape changes. A house (physical structure) is shown close to the roadway. In reality and as depicted on the larger-scaled (1:10,000) roadway drawing, the house appears left (or west) of the roadway. However, on the smaller-scaled (1:50,000) roadway depiction, the house is on the right (or east) of the roadway, forming a basic generalization of actual position and real-world physical shapes.

Decisions concerning the use of geographic data and their suitability for any given application are paramount to the validity of the geospatial system. Selecting an adequate scale for a project is a balance between positional accuracy and the amount of geographic data processing. For instance, take the more generalized road and apply it to a tourism booklet's map. Geographic data requirements are minimized. Additionally, the effect of the rounded road is perhaps negligible considering the native generality of the tourism application. However, take that same geographic data with the same "rounded" characteristics and apply it to an emergency vehicle deployment routing application. That mistake in judgment may have taken someone's life! The depicted geographic data at a 1:50,000 scale are generalized beyond minimum suitability for the emergency deployment application. The vehicle drivers are looking for a specific house on the left that is actually on the right! A 1:10,000 scale would be a more suitable level of generalization for ensuring the system's integrity for positional accuracy and left-side/right-side roadway events.

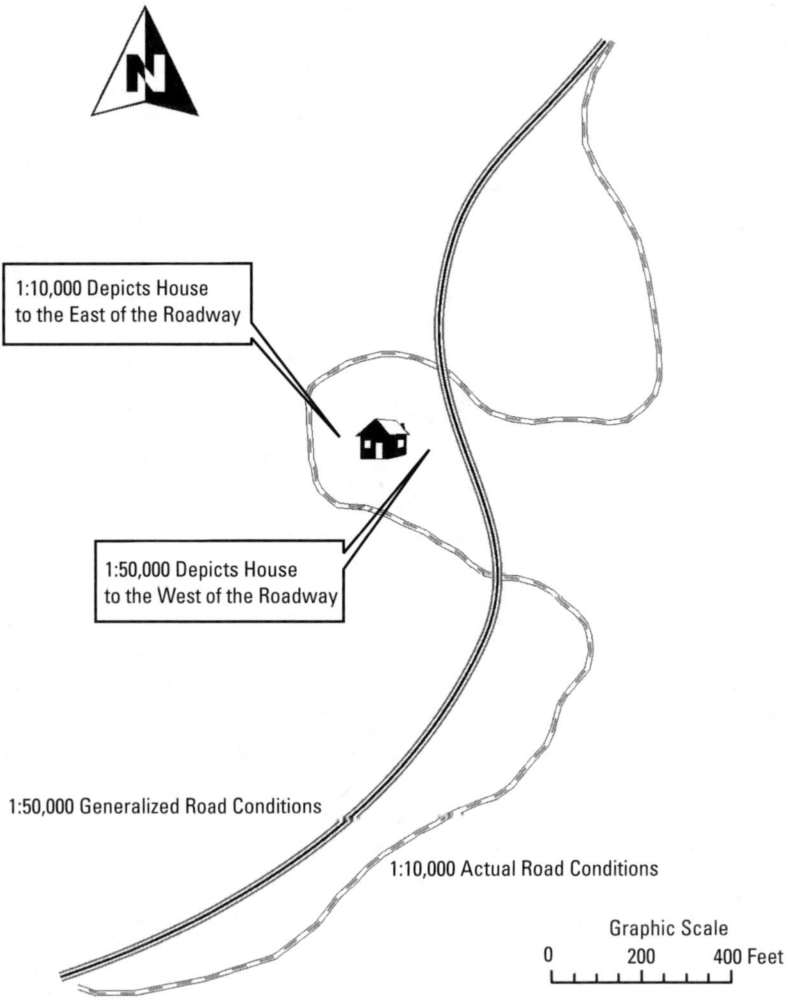

Figure 5.4 The "house on the wrong side of the road" problem.

6

The Ellipsoid

Throughout the early years of geodetic understanding, some of the most important components of geodesy were realized through innovative thinking, physical science, and debate. The spherical Earth-flat Earth debate previously discussed initiated the science of geodesy and spawned numerous attempts at defining the circumference of the Earth. By the seventeenth century, scientists, mathematicians, and others had begun to realize that the Earth wasn't quite perfectly rounded as first believed. New theories soon emerged out of the scientific communities, and the quest for the true Earth model began. This quest ended with the ellipsoid and subsequently recast physical science and geodesy forever.

Today's precision GIS is built upon the advances in modern geodesy and the finely tuned knowledge of the Earth's true shape. With satellite imagery and modern physical science, there is now no question as to what the world looks like and what shape best represents the Earth—a nearly round object that bulges in the middle. To this end, ellipsoids are exceedingly important to GIS because they pose a near resemblance to the Earth's shape without the inherent complexities. In short, ellipsoids simplify GIS.

To truly understand the core GIS environment, we must first understand how we got from Eratosthenes's curved Earth and the pure sphere to an ellipsoid, a much closer physical representation (model) of the real Earth. We will discuss the physical parameters of the ellipsoid and how this elliptical Earth model is manipulated and referenced within GIS.

History of the Earth Model

From the time Eratosthenes calculated the Earth's circumference in the third century B.C., the Earth model for many took on the shape of a sphere. Generally, a sphere is a symmetrical three-dimensional shape, in which the distance (radius) from the surface to a common center point is identical anywhere on the object (i.e., height equals width). It can be thought as a circle rotated around an axis.

The sphere shape for the Earth model was the initial conclusion based on a combination of minimal information and poor measurement tools. It was the logical curved successor to the "flat Earth.. To its credit, it remained a favored Earth model from the third century B.C. to the seventeenth century A.D., when science made it possible to measure distances and gravity with precision. Telescopes were invented and became indispensable tools for early geodesists.

Soon, imperfections in the Earth's curvature became widely apparent with the recognition and measurement of dissimilarities in different locations. This dissimilarity invoked new theories concerning the Earth model. As a direct result, two predominant and disputing theories emerged from the geodetic community, namely from the French and the British communities. Figure 6.1 details the competing Earth model theories.

The story of the British-French rivalry concerning the Earth model is legendary and quite interesting. Both wanted to be correct in order to attain a level of international excellence and scientific supremacy. Leading the French were the Cassinis, a renowned family of geodesists. After taking measurements in France from Paris to Perpignan, Jean-Dominique Cassini and his son, Jacques

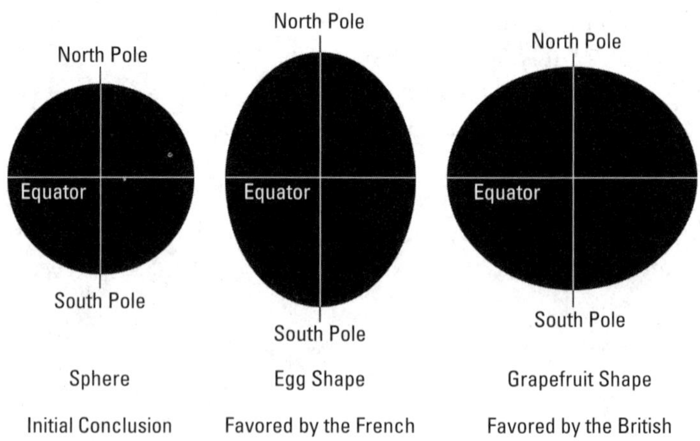

Figure 6.1 The predominant Earth models.

Cassini, were certain the Earth pointed toward the poles and flattened at the equator. As a result, the French model was a prolate ellipsoid and egg shaped.

The British took a different approach. They were armed with Isaac Newton's then recently released gravitational theory published in Newton's *The Principia.* The British were equally certain that the Earth was flattened but that it was occurring at the poles. The British model was an oblate ellipsoid—grapefruit shaped and bulging in the middle.

The competition became serious as international interest grew. In fact, the contrasting theories invoked an international debate with the Earth *elongators* favoring the French model and the Earth *flatteners* favoring the British model. In an effort to end the dispute and determine who was truly correct, the French Academy of Sciences devised a plan that involved measuring a 1° arc in the North and on the equator. If the arc in the North was shorter than the arc on the equator, then the French would be correct. If the North arc was longer, then the British would be correct.

The French Academy of Sciences sent out two teams to make 1° arc measurements in Peru and Lapland. Peru was a reachable point near the Equator and Lapland was reachable and near the North Pole. This meridian arc measurement expedition started in April 1735 as the first team was deployed to Peru. In April 1736, just two months shy of the first team reaching the site in Peru, the second team was deployed to Lapland.

By August 1737, the Lapland team was already finished and back in Paris. The Peru team, on the other hand, had a much harder time with its meridian arc measurement. Due to the extraordinarily rough terrain in the Andes mountains, the team did not complete its measurements until 1743 and did not return to Paris until February 1745. Incredibly, the French Academy of Sciences' resolution took nearly 10 years to complete. In the end, the British were proved correct and the Earth model took on its new form: a grapefruit-shaped oblate ellipsoid.

Shape of the Earth

The ellipsoid, also referred to as a *spheroid*, is a much better approximation for the shape of the Earth than the sphere; the poles are slightly flattened and the equator bulges. Unlike the sphere, the ellipsoid can handle dissimilar dimensions. GIS relies heavily upon geographic positioning and the Earth model; therefore, the closer a model comes to the actual surface of the Earth, the better for geographic positioning. A sphere proves adequate along the equator (equatorial plane) but fails at locations closer to the poles. For this reason a GIS based upon a sphere model would produce flawed information products for most locations throughout the world, especially within the arctic and antarctic circles.

This Earth flattening defines our contemporary understanding of the shape of the Earth. Pull out a table globe and examine its features. Although difficult to immediately recognize on a globe, notice how the nearly spherical shape that ancient theorists surmised as "perfect" slightly bulges at the equator. Now imagine placing an invisible measuring stick through the core from the north pole to the south pole. It would measure 12,713,505 meters. Imagine placing another measuring stick through the Earth's core from one point on the equator to another. That stick would measure 12,756,274 meters. This relatively tiny difference in measurements causes the Earth's subtle bulge and defines the nature of its flattening.

Figure 6.2 offers a bird's-eye look at ellipse flattening as a whole. As you can see, a circle has zero flattening because the horizontal and vertical measurements are equal. When the measurements differ, that is, when the horizontal measurement is larger than the vertical measurement, ellipse flattening increases. Alternately, a larger vertical measurement defines an ellipse elongation.

Figure 6.2 illustrates flattening with large, visually dramatic flattening scenarios. However, in the case of the Earth, the north-to-south measurement is only slightly shorter than the east-to-west measurement and has a small flattening of approximately 1/300.

What makes the ellipsoidal model useful in GIS and other real-life applications (i.e., GPS systems, military artillery, science, and so forth) is its ability to

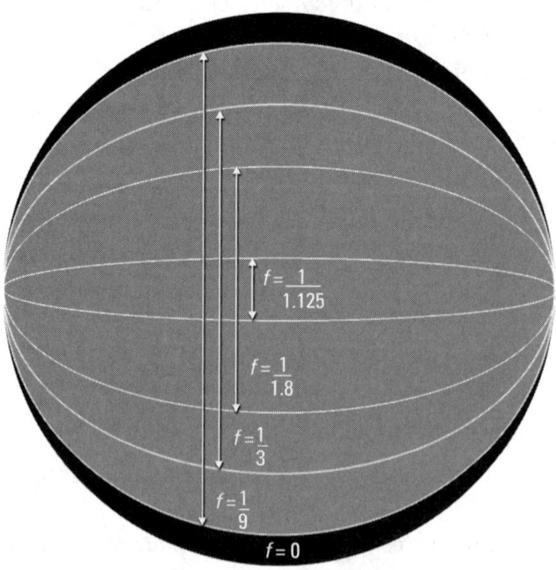

Figure 6.2 Various levels of flattening.

handle the Earth's flattening with ease and within the simple constructs of a relatively smooth shape. Now, the Earth is hardly smooth. We have mountains, plains, and valleys on land and deep, cavernous terrain within the oceans. We have cities that are below sea level (such as New Orleans, Louisiana) and others many miles above sea level (such as Katmandu in Nepal).

You might be thinking, how could the Earth's various terrains be served adequately by a smooth ellipsoid? Well, let us look at this question on a global level. The highest feature on Earth is Mount Everest at about 9 km above sea level. The lowest feature is the Pacific Ocean's Marianas Trench at roughly 11 km below sea level. The topographic height variation from highest to lowest is 20 km. Given that the shortest core-to-crust radius is 6,356.7 km, this topographic height variation would provide a minuscule surface change and is therefore negligible for Earth-modeling purposes.

We can liken these relatively subtle topographical height variations to the dimples on a golf ball; the overall shape is spherical, but the actual surface varies slightly in height. For ease, we use a smooth ellipsoid as the basis for calculations and geographic positioning. The difference between a smooth ellipsoid and one with topographic feature heights is negligible for many base-level calculations and for Earth referencing in GIS. We will look at these topographical height variations in greater detail in upcoming chapters.

Mathematical Model of the Earth

To adequately represent the shape of the Earth in scientific and real-life applications, a calculatable, formula-driven figure of the Earth is needed. Let us go to the basics of an ellipsoid to build a mathematical figure of the Earth. In its most basic sense, an Earth ellipsoid is a *flattened* sphere that bulges in the middle and has two imaginary lines traversing its core: one from north to south and one from east to west. Each imaginary line is called an *axis*.

The ellipsoid's flattening causes two unequal axes: a longer axis and a shorter axis. The north-to-south axis through the Earth's core is the shorter axis and, as such, is called the *minor axis* or *polar axis*. The east-to-west axis through the Earth's core is longer and is called the *major axis* or *equatorial axis*. Figure 6.3 illustrates a three-dimensional model and highlights the ellipsoid's flattening.

The Earth's ellipsoid is an ellipse rotated upon its minor axis, which is functionally called the *axis of rotation* or *axis of revolution*. The purpose of a mathematical model is to simplify calculations. Since an ellipse is a two-dimensional shape and an ellipsoid is a more complex, three-dimensional object, we utilize the figure of an ellipse to achieve this mathematical simplicity. Figure 6.4 details this two-dimensional mathematical model of the Earth.

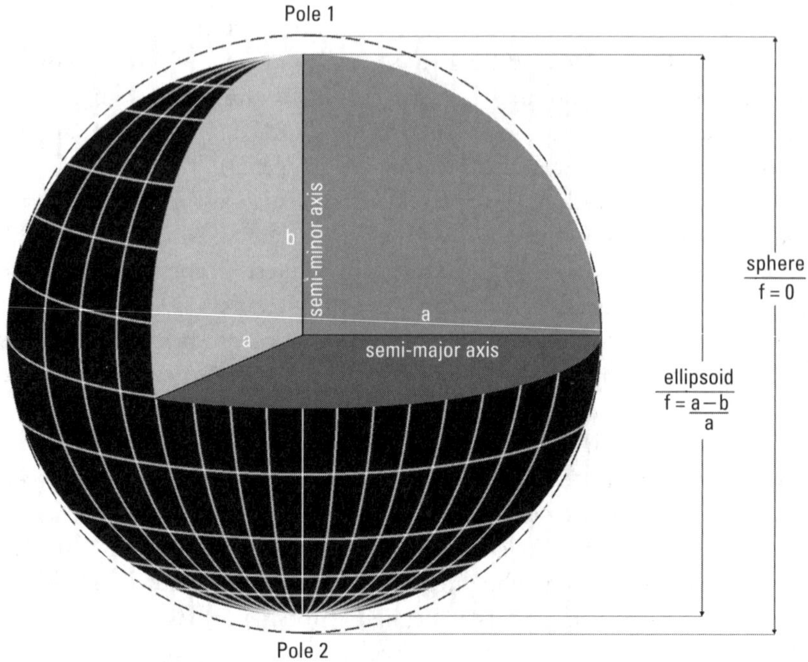

Figure 6.3 The three-dimensional oblate Earth.

Due to the Earth's symmetry and to minimize complexity, the mathematical model targets one quadrant in the ellipse and imparts six key parameters for calculation. The six parameters are the semimajor axis, the semiminor axis, flattening, inverse flattening, eccentricity, and second eccentricity. Let us go through each one while using Figures 6.3 and 6.4 for visualization.

The semimajor axis is an equatorial radius and is defined as one-half of the major axis. Alternately, the semiminor axis is a polar radius and is defined as one-half of the minor axis. Both hold major roles within the mathematical model and are represented by *a* and *b*, respectively. As we will see in the next two chapters, both the semimajor and semiminor axes are instrumental in defining accurate GIS point locations.

We have already discussed flattening but have not fully addressed its calculation. In brief, the flattening of the ellipse is directly related to the differences in both the semimajor and semiminor axes. It is represented by the formula

$$\mathit{Flattening}\ f = \frac{a - b}{a}$$

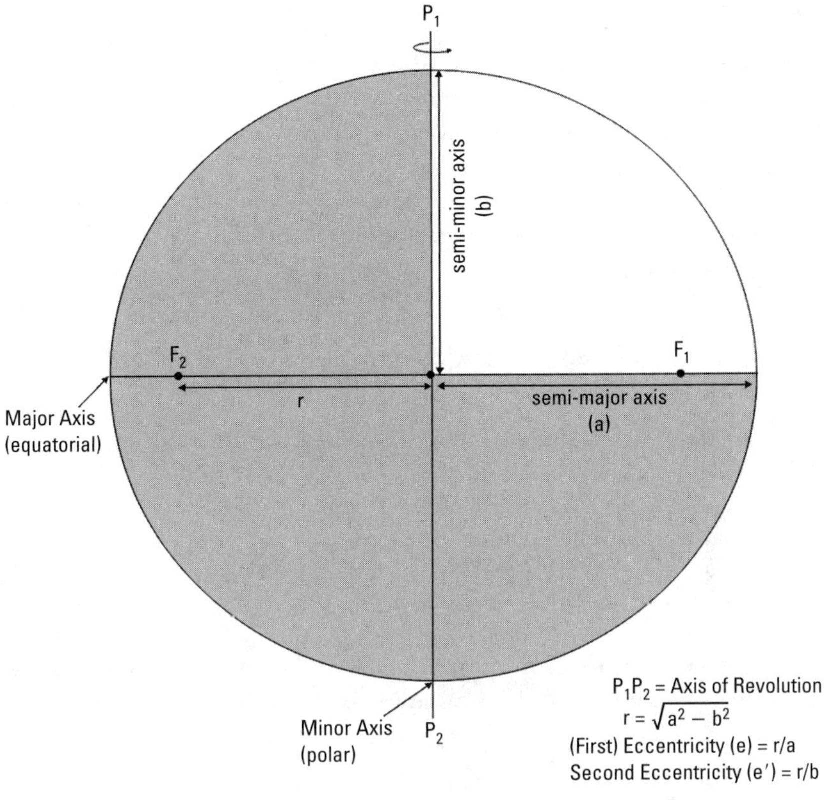

Figure 6.4 The two-dimensional, six-parameter mathematical model of the Earth.

Typical of this geodetic application, ellipse (and ellipsoid) flattening is a very small number holding many decimal places. In GIS, science, and mathematics, flattening results can be onerous to transcribe from tables and presents an opportunity for human error. To minimize error and help simplify conveyance, the Earth's flattening is sometimes displayed as a reciprocal called the *inverse flattening*. For instance, a flattening of 0.003389831 (or 1/295) can be portrayed as an inverse flattening of 295. As you can understand, the inverse flattening is much easier on the eyes and even easier to incorporate into dense calculations.

The formula for inverse flattening is straightforward, represented as:

$$\textit{Inverse flattening } f^{-1} = \frac{1}{f} \text{ or } \frac{a}{a-b}$$

The final two parameters, namely eccentricity and second eccentricity, have more to do with the mathematical nature of the ellipse than its geographic use. Both eccentricities uniquely characterize the shape of the ellipse through a relational measurement for the degree of flattening.

Eccentricity e is the first eccentricity, whereby the descriptor *first* is typically implied. It measures the degree of flattening as the relationship between the focal radius, or radius from the center point to the foci (i.e., F_1, F_2), and the semimajor axis. It is calculated through the formula

$$(First)\ Eccentricity\ e = \frac{Focal\ radius\ r}{Semimajor\ axis\ a}$$

In any ellipse, the focal radius $r = \sqrt{a^2 - b^2}$; therefore,

$$(First)\ Eccentricity\ e = \frac{\sqrt{a^2 - b^2}}{a}$$

Second eccentricity e' measures the degree of flattening as a relationship between the focal radius and the semiminor axis. The formula for second eccentricity is

$$Second\ \ Eccentricity\ e' = \frac{Focal\ radius\ r}{Semiminor\ axis\ b} = \frac{\sqrt{a^2 - b^2}}{b}$$

Eccentricity (first and second) is constant throughout the entire ellipse and ranges from 0 to 1, or more simply from an ellipsoid with no flattening (circle) to one totally flat (line). Since a circle has no flattening, its eccentricity equals zero ($e = 0$). As e becomes larger and approaches 1, the ellipse becomes flatter until it is a straight line ($e = 1$).

Figure 6.5 illustrates this proportional relationship and endorses the eccentricity rule of thumb: The closer eccentricity is to 1, the flatter the ellipse.

Typical Earth Shape Calculation

To understand how the mathematical model is used for typical Earth shape calculations in GIS, let us calculate the Earth's six key parameters using the derived mathematical model as illustrated in Figure 6.4 and the standard measurements of the Earth's axes from the World Geodetic System 1984 (WGS84).

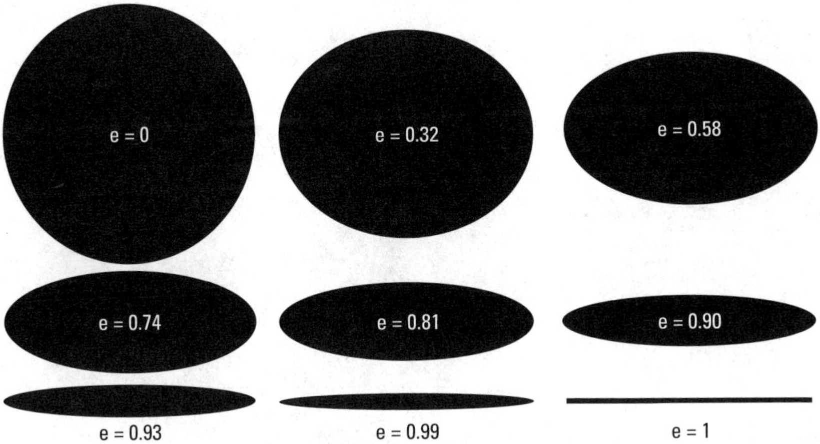

Figure 6.5 The constant eccentricity of an ellipse. (*Source:* D. Downing, *Dictionary of Mathematics Terms*, 2nd ed., Hauppauge, NY: Barron's Educational Series, 1995, after Figure 47.)

WGS84 Constants

$$\textit{Major axis} = 12{,}756{,}274 \text{ meters}$$

$$\textit{Minor axis} = 12{,}713{,}504.628 \text{ meters}$$

$$\begin{aligned}\textit{Semimajor axis } a &= \tfrac{1}{2}\textit{ Major axis}\\ &= \tfrac{1}{2}(12{,}759{,}274)\\ &= 6{,}378{,}137 \text{ meters}\end{aligned}$$

$$\begin{aligned}\textit{Semiminor axis } b &= \tfrac{1}{2}\textit{ Minor axis}\\ &= \tfrac{1}{2}(12{,}713{,}504.628)\\ &= 6{,}356{,}752.314 \text{ meters}\end{aligned}$$

$$\begin{aligned}\textit{Flattening } f &= \frac{a-b}{a}\\ &= \frac{6{,}378{,}137 - 6{,}356{,}752.314}{6{,}378{,}137}\\ &= \frac{21{,}384{,}686}{6{,}378{,}137}\\ &= 0.00335811\end{aligned}$$

$$\textit{Inverse flattening } f^{-1} = \frac{1}{f}$$
$$= \frac{1}{0.003352811}$$
$$= 298.257$$

$$\textit{Eccentricity } e = \frac{r}{a} = \frac{\sqrt{a^2 - b^2}}{a}$$
$$= \frac{\sqrt{6{,}378{,}137^2 - 6{,}356{,}752.314^2}}{6{,}378{,}137}$$
$$= \frac{\sqrt{272{,}331{,}609{,}224.645404}}{6{,}378{,}137}$$
$$= \frac{521{,}854.0114}{6{,}378{,}137}$$
$$= 0.081819191$$

$$\textit{Second eccentricity } e' = \frac{r}{b} = \frac{\sqrt{a^2 - b^2}}{b}$$
$$= \frac{\sqrt{6{,}378{,}137^2 - 6{,}356{,}752.314^2}}{6{,}356{,}752.314}$$
$$= \frac{\sqrt{272{,}331{,}609{,}224.645404}}{6{,}356{,}752.314}$$
$$= \frac{521{,}854.0114}{6{,}356{,}752.314}$$
$$= 0.082094438$$

Reference Ellipsoids

Before leaving the subject of the ellipsoid, we must define its function within a GIS. Basically, every GIS and geographic information product is based upon a reference ellipsoid, which is defined as a standard ellipsoid with proven and measured parameters. The reference ellipsoid enables GIS to define point locations with regional accuracy.

Throughout the years, numerous reference ellipsoids have been developed by various geodesists with measurements taken at different source points on Earth. These reference ellipsoids are only slightly dissimilar. Because of this differentiation in source data locations, certain reference ellipsoids work better than others for certain applications and for certain regions. For example, the reference ellipsoid Bessel 1841 is better suited for European GIS products than American GIS products and will produce more precise results due to its European-based measurement site.

Unlike Bessel 1841 and other major area-specific ellipsoids, WGS84 is much more universal and, as a result, is one of the most widely used reference ellipsoids available. It can be used for any location on Earth with a high degree of precision and flexibility and is backed by satellite measurements.

Appendix A, entitled "Reference Ellipsoid Parameters," lists many of the key reference ellipsoids available to GIS users. The appendix provides their associated equatorial radius, polar radius, and inverse flattening parameters, which are all necessary for proper designation in GIS. Since all geographic information products are built upon a reference ellipsoid, we will frequently refer to Appendix A throughout this book.

7

The Horizontal Datum

During the "Shock and Awe" military mission in Iraq, the U.S. armed forces were responsible for destroying enemy targets with missiles. The challenge was that these targets were in the heart of downtown Baghdad, near schools, mosques, and neighborhoods. The accuracy of missiles had to be extremely precise so that only the target was destroyed and the surrounding innocent people were not killed. Using standard ellipsoids to model the Earth helped bring the science of geodesy into a manageable and useful format for various applications. However, launching those missiles using positions on a standard ellipsoid would have been catastrophic, since positional accuracy would have been generalized, and thus the target's position would have been generalized. To maintain precision, U.S. missile systems implemented GPS-based targeting systems for the highest level of accuracy and to minimize collateral damage.

Sometimes standard ellipsoids provide enough accuracy and prove sufficient for small- and large-scale GIS applications. However, similar to the military missile systems, often a greater degree of precision and positional accuracy is needed. To help bridge the standard ellipsoid to a more exact Earth position, a precise reference point called a *geodetic datum* is used. Incidentally, specific geodetic datums are used to manipulate GPS properly and help guide missile targeting systems.

Let us examine these precision-enhancing ellipsoid references to obtain a clearer understanding of their importance and implementation within GIS. The complete discussion of datums is somewhat lengthy. In an effort to simplify and help organize the explanation of the topic, I have divided it into two logical parts: the horizontal control datums and the vertical control datums. This chapter focuses on datum theory, as well as horizontal control datums. The next chapter completes the discussion with a detailed examination of vertical control datums.

What Is a Geodetic Datum?

By the purest sense of the word, a datum is any set of numeric or geometric parameters used to accurately measure or define another quantity. More specifically related to GIS, a geodetic datum is a "curved Earth" reference model that associates a geodetic reference ellipsoid (i.e., "parameters") to a coordinate system (i.e., "another quantity").

As previously discussed, a geodetic reference ellipsoid is attributed with certain characteristic parameters, such as the semimajor axis *a*, the semiminor axis *b*, and flattening *f*, while a coordinate system defines a geodetic space through orientation, position, and scale. Both are brought together with a geodetic datum, which, in itself, is purely a mathematical Earth model. Given this mathematical union it can be determined that any reference ellipsoid can support an infinite number of datums; however, each datum can only be defined by one ellipsoid.

Geodetic datums are generally classified as either a geocentric datum or a local geodetic datum. A *geocentric datum* is globally centered and adequately approximates the Earth's size and shape as a whole. The center of the reference ellipsoid coincides with the Earth's center of mass. See Figure 7.1.

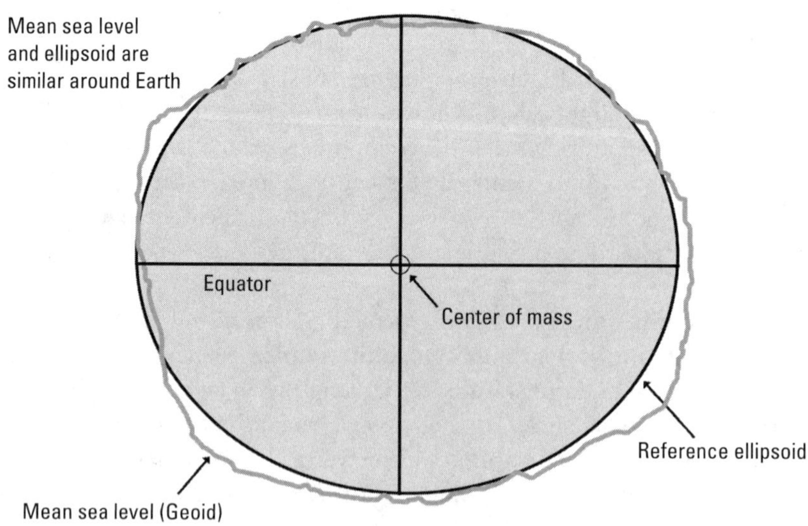

Figure 7.1 The geocentric datum. (*After:* [1], diagram 6.)

A geocentric datum is not designed to best characterize a particular area but instead is best used when applications are global in nature. GPS, for example, utilizes the common geocentric datum WGS84.

A *local geodetic datum* is used to characterize a particular region where the reference ellipsoid and the Earth's shape coincide. The center of the ellipsoid is often located away from the Earth's center of mass, thereby providing a poor representation of the Earth as a whole. Figure 7.2 illustrates how the ellipsoid's surface and Earth's surface coincide within a particular location and diverge elsewhere.

Both figures depict the Earth's mass as a geoid or mean sea level model. This is a typical convention with geodetic datums, especially when defining the height of the Earth from its core. Since we are detailing the relationship between the Earth's center of mass and the center of the reference ellipsoid, as well as the surfaces of both these shapes, the geoid proves to be a fair Earth shape parameter. We will examine the true function of the geoid in the next chapter.

Primary Types of Geodetic Datums

The classification of geodetic datums offers a degree of user control as it relates to which datum suits the application or location best. For instance, you would never choose NAD27, a North American datum, to map the city of Tokyo in Japan. The Tokyo datum has been devised specifically to handle that terrestrial location. This form of user control surrounds the act of selecting a

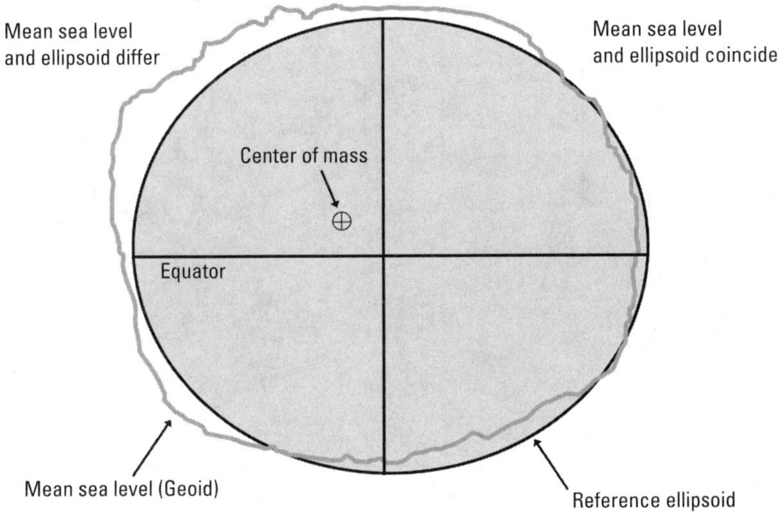

Figure 7.2 The local geodetic datum. (*After:* [1], diagram 5.)

location-specific datum based on individual characteristics, rather than manipulating a more general datum.

Geodetic datums empower the user with a referenced-based geographic coordinate system within which accurate point positions can be determined. In a two-dimensional space, point positions are defined by a coordinate pair (*x*, *y*) on a uniform plane. Think of a single point on an *x*-*y* grid drawn on a piece of paper. Every *x*-*y* positional coordinate on that paper is at a uniform height (i.e., uniform plane). In the real three-dimensional world, height (*z*) is as important a distinguisher as the *x* and *y* coordinates. The three-dimensional point positions involve a horizontal component (*x* and *y* coordinates) and a vertical component (*z* coordinate).

At any given point on the Earth's surface there exists a horizontal level that runs parallel with the surface (ground) and a vertical level that runs perpendicular to the surface (ground). Refer to Figure 7.3.

Imagine holding a plumb line or a weighted string so that it remains suspended in the air aiming toward the ground. The imaginary line running through the string and weight into the ground provides that point position's vertical level. Likewise, a glass or cylinder containing water and placed on the ground presents the horizontal level for that position. Since horizontal and vertical levels exist for every point position, positional coordinates can be adjusted as needed along these planes.

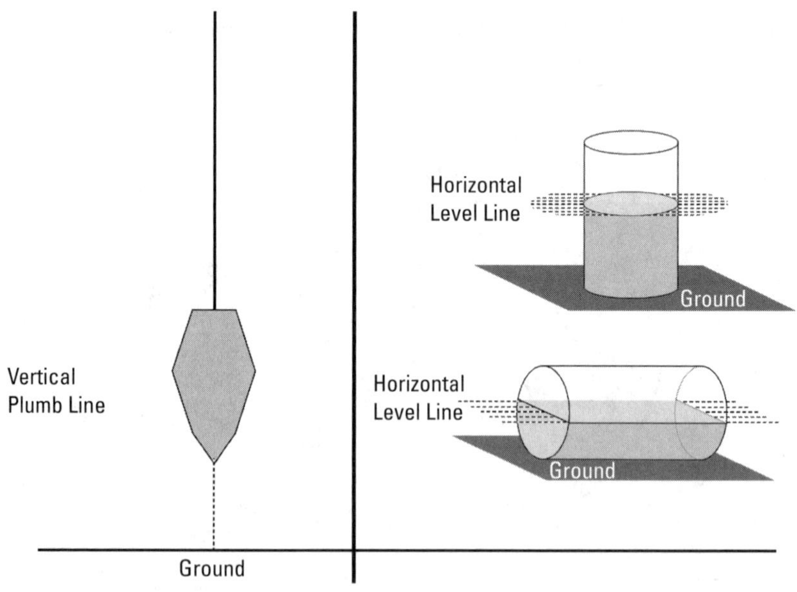

Figure 7.3 The vertical plumb line and the horizontal water line.

Geodetic datums, both local and geocentric, offer the opportunity to manage these horizontal and vertical levels for greater positional accuracy. This leads us into the two primary types of geodetic datums: horizontal control datums and vertical control datums.

A *horizontal control datum* (or *horizontal datum*) is used to manipulate and set a position in the *x* and *y* directions. Horizontal datums are often defined by a reference ellipsoid and a coordinate origin. Alternately, a *vertical control datum* (or *vertical datum*) is used to manipulate and set a positional height in the *z* direction. Vertical datums provide a level from which positional heights can be determined.

Both primary types of geodetic datums provide a favorable degree of positional control. As mentioned earlier, the remainder of this chapter will examine the standard types of horizontal datums and various methods to transform between two horizontal datums.

Standard Horizontal Datums

As we already understand, horizontal datums take on both geocentric and local geodetic forms. Each major province or country has its own standard datum or simply a datum of choice. Standard horizontal datums cover large geographic areas. They are used throughout the world for simplicity and local datum reference. We will briefly examine five commonly used horizontal datums and reference systems for the Earth model.

World Geodetic System 1984

The World Geodetic System 1984 (WGS84) is considered a global datum that defines an Earth-fixed global reference frame for the Earth. WGS84 is based upon the ellipsoidal parameters of the Geodetic Reference System of 1980. WGS84 is used for GPS satellites and is defined by a gravity model of the Earth.

WGS is a geocentric datum that made its first appearance in 1960. A product of the U.S. Defense Mapping Agency (DMA) for mapping and charting, the WGS has evolved from the primitive satellite measurements of 1960 (WGS60) to the sensitive Doppler satellite data and satellite data altimetry used in the latest and highly accurate WGS84. With its satellite-driven precision, WGS84 has proved to be a satisfactory reference datum for nearly every Earth location.

Given its Earth-fixed nature, WGS84 offers a standard reference ellipsoid with base parameters being:

$$\textit{Semimajor axis } a = 6{,}378{,}137 \text{ meters}$$
$$\textit{Semiminor axis } b = 6{,}356{,}752.314 \text{ meters}$$
$$\textit{Flattening } f = 0.003352811$$
$$\textit{Inverse flattening } 1/f = 298.2572236$$

Due to its general accuracy worldwide, WGS84 is a highly used, worldwide datum (and reference ellipsoid). In fact, due to its global appeal, many applications require local datums to be converted to WGS84. The transformation parameters provided with the worldwide geodetic datums in Appendix B are conversion factors between the particular local datum and WGS84.

Incidentally, it is anticipated that WGS84 will remain valid until 2010, when it will be revised.

International Terrestrial Reference Frame

Another major global reference datum is the properly named International Terrestrial Reference Frame (ITRF). This reference system is used for a large selection of astronomical, geophysical, and geodetic analysis and information products worldwide, although it is typically more popular in Europe.

The goal of this geodetic datum is the realization of a terrestrial coordinate system through the creation and implementation of a terrestrial reference frame. Positions on the surface of the Earth are indeterminate observations or quantities without a terrestrial reference frame. Thus, the ITRF enables precise positional coordinates that are linked to the terrestrial landscape of Earth. This terrestrial landscape is referenced through an International Terrestrial Reference System (ITRS), which is a dynamic spatial referencing system that adheres to the Earth's rotation and geophysical changes.

The ITRF has proved to be a uniquely useful and precise tool for both the geodesist and GIS user alike. The ITRF maintains precision through use of modern geodesy, GPS, satellite laser ranging (SLR), lunar laser ranging (LLR), very long baseline interferometry (VLBI), and other analysis tools. Additionally, precision is maintained with the aid of measurement for more than 800 observation and control stations worldwide. These observation and control stations are typically located at small distances from major fault lines or areas of seismic activity. The resultant product, namely the ITRF, is a highly precise datum that simulates the present-day terrestrial Earth and captures the minute changes in the Earth that other geocentric datums cannot capture.

The ITRF's base parameters for its reference ellipsoid are identical to WGS84 down to the 10-cm level. This level of accuracy is precise enough to say that WGS84 and ITRF share the same ellipsoid parameters. In addition, ITRF has no transformation factors to WGS84. As a result, ITRF coordinates can be expressed as WGS84 coordinates at a 10-cm level.

The first adequate version of ITRF came out in 1994. It was revised in 1996, again in 1997, and finally in 2000. At the time of this writing, the latest form is ITRF2000, although plans are set for an ITRF2005.

Geodetic Reference System of 1980

The Geodetic Reference System of 1980 (GRS80) is another widely used geodetic reference system that consists of a global reference ellipsoid and a gravity field model. The GRS80 model provides a geocentric reference ellipsoid used by the ITRF geodetic network. GRS80 supersedes GRS67. Both were developed and are maintained by the International Association of Geodesy (IAG), a division of the International Union of Geodesy and Geophysics (IUGG). The GRS80 model offers a simple yet accurate framework for the definition of a worldwide geodetic datum and the determination of the normal gravity field for the Earth.

The GRS80 ellipsoid encompasses the entire mass of the Earth, which goes beyond the terrestrial mass and includes the atmosphere. The inclusion of the Earth's atmosphere into the ellipsoid measurements was an innovation first introduced in the GRS80 model.

The GRS80 reference system is used by GPS through a standardized realization of the system: WGS84. Given this connection, GRS80 offers a standard reference ellipsoid with similar base parameters.

$$\textit{Semimajor axis } a = 6{,}378{,}137 \text{ meters}$$
$$\textit{Semiminor axis } b = 6{,}356{,}752.314 \text{ meters}$$
$$\textit{Flattening } f = 0.003352811$$
$$\textit{Inverse flattening } 1/f = 298.2572236$$

As mentioned earlier, GRS80 takes into account the Earth's atmosphere, which is conceptually a uniform, condensed layer of atmosphere on the surface of the ellipsoid. In the GRS gravity field model, the variation in density for this condensed atmospheric layer is ignored so that the geocentric gravitational field can be efficiently calculated. The GRS80 gravity field model is defined through the following physical constants:

$$\textit{Geocentric gravitational constant } GM = 3{,}986{,}005 \times 10^{8} \text{ m}^3/\text{s}^2$$
$$\textit{Dynamic form factor } J_2 = 108{,}263 \times 10^{-8}$$
$$\textit{Angular velocity of the Earth } \omega = 7{,}292{,}115 \times 10^{-11} \text{ rad/s}$$

The GRS80 geocentric model is a reference system through which other geocentric and local geodetic datums are based. Its worldwide accuracy has enabled GRS80 to remain a geodetic datum reference of choice for many regions. For instance, the government of Australia used the GRS80 geocentric ellipsoid to define GDA94, a popular Australian local datum.

North American Datums of 1927 and 1983

The North American Datums of 1927 and 1983 were established to provide accurate positional coordinates and increased horizontal control for any location within North America. By the end of 1926, the original horizontal reference system for the United States (North American Datum) was drastically outdated and imprecise, and the U.S. Coast and Geodetic Survey called for a revision to the standard datum. In 1927, the North American Datum of 1927 (NAD27) was developed to include the then newly measured meridian arcs of North America. The new datum was based on the Clarke ellipsoid of 1866. NAD27 proved to be a more accurate geodetic local datum that is still used today.

Similar to the original North American Datum, NAD27 is referenced through a base triangulation station in Meades Ranch, Kansas. This reference station is close to the geographic center of the United States, which helps NAD27 minimize differences between astronomic and geodetic latitudes and longitudes.

The development of GRS80 brought forth the many shortcomings of NAD27 and showed the need for a revised and more accurate North American Datum. The North American Datum of 1983 (NAD83) was subsequently developed through an international effort between the U.S. National Geodetic Survey, the Geodetic Survey of Canada, and the Danish Geodetic Institute.

The resulting NAD83 is a geocentric datum based upon the GRS80 Earth model that uses three-dimensional techniques to establish horizontal control. These (x, y, z) observations are facilitated through the implementation of high-precision measurement tools, namely GPS, Doppler satellite, and VLBI. In fact, precision of these measurement tools is within 2 meters.

As a geocentric datum, NAD83 uses the Earth's mean center of mass as a datum reference, whereas NAD27 uses a reference on the Earth's surface (the Meades Ranch station). Adding to NAD83's importance is its compatibility with the WGS84 (GPS) network. In fact, the differences between the two models are negligible. For the purposes of GISs, these datums are effectively equivalent.

The differences between coordinates in NAD27 and NAD83 range from 200 to 300 feet in the western United States to tens of feet in central and eastern U.S. regions. Coordinate conversion between NAD27 and NAD83 can be done using GIS software or geographic calculators or through the horizontal datum transformation methods discussed later in this chapter. Eventually, there will be only one North American Datum, since NAD27 is gradually being superseded by NAD83.

The reference ellipsoid parameters for the North American Datum of 1927 are:

$$\textit{Semimajor axis } a = 6{,}378{,}206.4 \text{ meters}$$
$$\textit{Semiminor axis } b = 6{,}356{,}758.8 \text{ meters}$$
$$\textit{Flattening } f = 0.003390075$$
$$\textit{Inverse flattening } 1/f = 294.9786982$$

The reference ellipsoid parameters for the North American Datum of 1983 are:

$$\textit{Semimajor axis } a = 6{,}378{,}137 \text{ meters}$$
$$\textit{Semiminor axis } b = 6{,}356{,}752.314 \text{ meters}$$
$$\textit{Flattening } f = 0.003352811$$
$$\textit{Inverse flattening } 1/f = 298.2572236$$

European Datum 1950

During World War II, the accuracy and compatibility of latitude and longitude positions in Western Europe became an issue. Battles fought on the borders of Germany, France, Belgium, and the Netherlands proved that datum inconsistencies existed, while the mapping of these countries aired several incongruities. The common consensus was that a consistent mapping datum was needed in Western Europe. As such, the European Datum 1950 (ED50) was created after the war.

ED50 is a geodetic datum that proves to be highly accurate in Western European countries. ED50 has abolished the incongruities and incompatibility issues that were faced during the many battles of World War II, while providing a tool to connect the existing and individual country datums.

ED50 is based upon the International 1924 ellipsoid (known as the Hayford ellipsoid) and offers the following base parameters:

$$\textit{Semimajor axis } a = 6{,}378{,}388 \text{ meters}$$
$$\textit{Semiminor axis } b = 6{,}356{,}911.946 \text{ meters}$$
$$\textit{Flattening } f = 0.003367003$$
$$\textit{Inverse flattening } 1/f = 297$$

Similar to other pre-1960 datums, ED50 is referenced through a base triangulation station. This European geodetic datum is referenced at Frauenkirche of Munich in Southern Germany, close to the center of Western Europe's geodetic networks during the Cold War. Before the implementation of GRS80 and WGS84, the ED1950 was widely used around the world. Today, it is still used by default in most of Western Europe except in countries with their own datums, such as Great Britain, Poland, Ireland, Hungary, Sweden, and Switzerland.

Horizontal Datum Transformation Models

With the multitude of horizontal geodetic datums available for any given region, the need for datum consistency is essential for correct and accurate results. Combining coordinates from different datums will produce incorrect results and can even put users in jeopardy. That being stated, often GIS users or others need to transform the coordinates from one geodetic datum to another for the sake of coordinate consistency and accuracy.

Imagine that you are trekking in the backlands of Malawi, Africa, with a region map based on the local datum (ARC1950) and realize you are moving beyond the boundaries of this map. You take out your GPS unit and quickly recognize it functions on a different datum (WGS84). To reach your destination and return safely, you will need to transform your positional coordinates from the local datum (ARC1950) into the GPS-friendly datum (namely WGS84). A datum transformation must be calculated not only to obtain your GPS coordinates but also to link the local map with the GPS unit.

This example is dramatic, but not unlikely. Geodetic datum transformations are common practice for GIS users who are working with existing coordinate data and different geodetic datums. In fact, datum transformation is the most utilized positional management function in core GISs. Geodetic network orientation is frequently mismatched between sets of geographic data and requires correction before properly coinciding in geographic space. A datum transformation provides such correction.

As you might have guessed, there are a handful of accurate transformation models from which to choose. Although many transformations can be calculated using software, the novice GIS user should understand the methods for calculating these transformations. Most applications of these methods seek to transform geographic data to the standard WGS84 from older or more obscure geodetic datums.

WGS84 serves as a common model for all geodetic networks and, therefore, acts as a common ground for many datum transformations. More specifically, many datums are first transformed into the common WGS84 and then transformed again into the desired datum. By expressing datum parameters as "shifts" to WGS84, the need to define all geodetic networks in terms of all other geodetic networks is alleviated. The worldwide datums table in Appendix B lists an assortment of local geodetic datums from around the world with their *x*, *y*, and *z* datum shifts to WGS84.

Let us examine the two most common geodetic datum transformation formulas: the Molodensky datum transformation method and the Bursa-Wolf datum transformation method.

The Molodensky Datum Transformation Method

The Molodensky (or Molodenski) datum transformation method is by far the most efficient and simplest datum transformation method available. Although sufficiently accurate, its simplicity makes this transformation less precise than others, stemming from the fact that it does not factor in rotation or scaling between datums. The Molodensky transformation is commonly called the *three-parameter transformation method* for its minimal formula involving three datum shift parameters between the two ellipsoids. This transformation method is used when the *X*, *Y*, and *Z* axes of both datums are known to be parallel and at the same scale. Further simplifying the calculation, the Molodensky method employs geodetic coordinates for the Earth's curvilinear surface, which alleviates any rectangular coordinate system conversion calculations.

As illustrated in Figure 7.4, the three geodetic datum shift parameters are defined as

1. Delta X (ΔX), the difference in geodetic longitude in meters between the current datum (#1) and the desired datum (#2);
2. Delta Y (ΔY), the difference in geodetic latitude in meters between the current datum (#1) and the desired datum (#2);
3. Delta Z (ΔZ), the difference in geodetic height in meters between the current datum (#1) and the desired datum (#2).

The Molodensky transformation formulas utilize the three datum shifts to calculate the desired datum coordinates. The formulas are as follows:

$$X_1 = X_2 + \Delta X$$

$$Y_1 = Y_2 + \Delta Y$$

$$Z_1 = Z_2 + \Delta Z$$

S. M. Molodensky first presented his datum transformation method in his classic title, *Methods for the Study of the External Gravitational Field and Figure of*

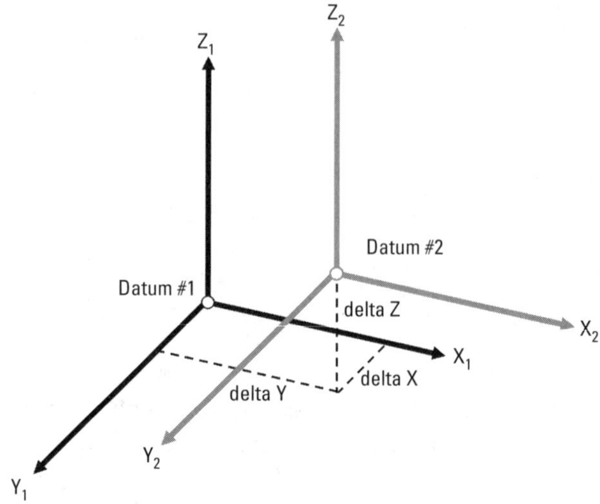

Figure 7.4 The Molodensky three-parameter datum transformation.

the Earth. His initial presentation involved a seven-parameter transformation method, which was further simplified to the current form: three parameters. In the end, the Molodensky datum transformation method retains adequate precision between the datums (within 5 meters) and proves to be a great geodetic tool for GIS.

The Bursa-Wolf Transformation Method

The Bursa-Wolf datum transformation method is a far more rigorous calculation than the three-parameter Molodensky formula. The Bursa-Wolf method is commonly called the *seven-parameter transformation method* for its lengthy formula involving three datum shift parameters, three datum rotations, and a scale correction factor. The Bursa-Wolf transformation is used when the *X*, *Y*, and *Z* axes of both datums are not parallel (axial rotation exists) and not at the same scale (scale change exists).

Similar to the Molodensky method, the Bursa-Wolf transformation method employs geodetic coordinates for the Earth's curvilinear surface, which, once again, alleviates any rectangular coordinate system conversion calculations. As illustrated in Figure 7.5, the seven datum parameters are defined as:

1. Delta X (ΔX), the difference in geodetic longitude in meters between the current datum (#1) and the desired datum (#2);

2. Delta Y (ΔY), the difference in geodetic latitude in meters between the current datum (#1) and the desired datum (#2);
3. Delta Z (ΔZ), the difference in geodetic height in meters between the current datum (#1) and the desired datum (#2);
4. X-axis rotation R_x, the rotation of the geodetic longitude in arc seconds from the current datum to the desired datum;
5. Y-axis rotation R_y, the rotation of the geodetic latitude in arc seconds from the current datum to the desired datum;
6. Z-axis rotation R_z, the rotation of the geodetic height in arc seconds from the current datum to the desired datum;
7. Scale change S_C, the scale difference between the current datum and the desired datum.

The Bursa-Wolf transformation formula utilizes all seven parameters to calculate the desired datum coordinates. The formula is as follows:

$$\begin{bmatrix} X_1 \\ Y_1 \\ Z_1 \end{bmatrix} = \begin{bmatrix} \Delta X \\ \Delta Y \\ \Delta Z \end{bmatrix} + S_c \begin{bmatrix} 1 & -R_z & R_y \\ R_z & 1 & -R_x \\ -R_y & R_x & 1 \end{bmatrix} \begin{bmatrix} X_2 \\ Y_2 \\ Z_2 \end{bmatrix}$$

The Bursa-Wolf transformation is a subset to the older Helmert seven-parameter transformation method. Both methods share the same formula, but they differ in their conventions for rotational polarity. The Helmert model (known also as *coordinate frame transformation* or *three-dimensional similarity transformation*) defines the rotation of the entire coordinate frame (i.e., all points on the axes). The Bursa-Wolf model (known also as a *position vector transformation*) defines the rotation of the point position with respect to a fixed coordinate frame. The conventions contrast one another and, thus, form two separate transformation methods. Compare Figure 7.5 (the Bursa-Wolf model) to Figure 7.6 (the coordinate frame transformation model).

The rotational polarity for the Bursa-Wolf model is positive clockwise around the datum 2 axes when viewed from the origin. Mathematically, the rotational polarity is reversed when viewing the origin from the axes. Whatever the convention used, the seven-parameter transformation model proves to be a much more accurate datum transformation method, with approximate accuracy held within 1 meter.

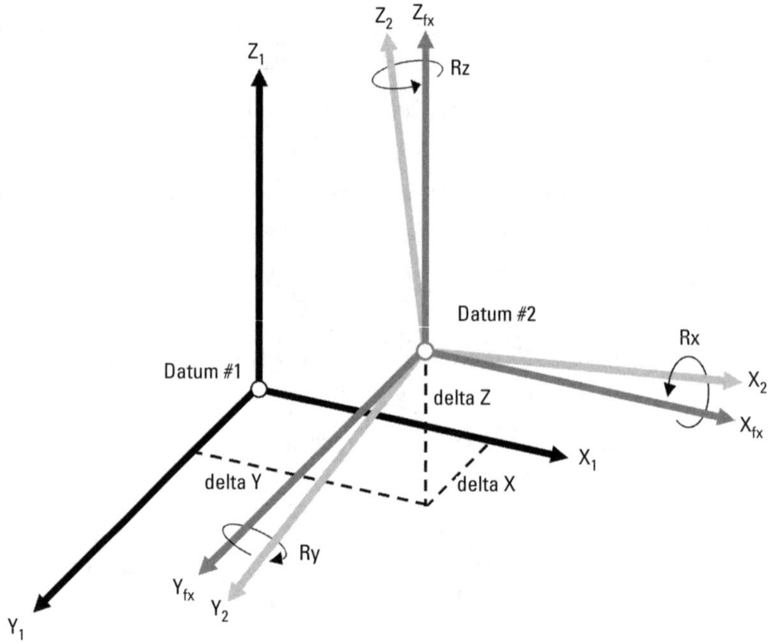

Figure 7.5 The Bursa-Wolf seven-parameter datum transformation.

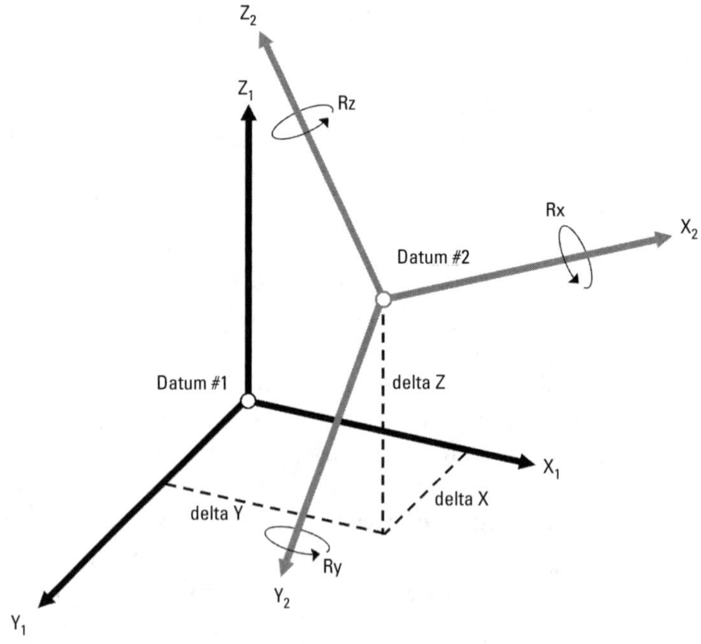

Figure 7.6 The coordinate frame transformation.

Horizontal Grid Transformation Methods

Another popular datum transformation type exists, called *horizontal grid transformation,* that enables discrete horizontal control within a datum. This transformation method is markedly different from both the Molodensky and Bursa-Wolf models. The latter models utilize a set of datum shift parameters that are attributed to a common source. Horizontal grid transformation methods use a set of field-based observations and statistical functions that describe a geodetic datum in a more comprehensive manner.

Adding to the complexity of the calculations but enhancing the accuracy of the results, horizontal grid transformations solve transformation through a mathematical function: bilinear gridded interpolation. This data interpolation technique facilitates the accurate transformation from older datum to newer, more accurate datum. Bilinear gridded interpolation is used by the U.S.-based North American Datum Conversion software and the Canadian National Transformation application. Additionally, this grid transformation technique has been adopted by the governments of Australia, New Zealand, and Great Britain for its high-accuracy horizontal control.

The standard horizontal grid transformation methods provide a greater level of horizontal datum control than that achieved using the Molodensky and Bursa-Wolf models. Adding to the better horizontal control is an enhancement in transformation accuracy, with grid transformations retaining an approximate accuracy within 0.1 to 0.3 meter. To reprise, the Bursa-Wolf accuracy was within approximately 1 meter and the Molodensky accuracy was within approximately 5 meters.

There are three very common horizontal grid transformation methods that warrant discussion: The North American Datum Conversion, the High Accuracy Reference Network, and the Canadian-Australian National Transformations. Let us examine these methods.

The North American Datum Conversion

The North American Datum Conversion (NADCON) is the U.S. federal standard for transforming latitude and longitude coordinates between NAD27 and NAD83. NADCON is a conversion utility that was first developed to facilitate transformations between the datums directly through known grid shifts, alleviating the need for datum parameters or scale changes.

A NADCON transformation is based upon a gridded set of control points that are geostatistically interpolated. NADCON grids are based on more than 150,000 of these horizontal control points. Transformations are based upon the grid shifts measured for the horizontal control point closest to the positional coordinate to be converted.

The NADCON program also provides datum correction for errors that exist in the NAD27 horizontal control datum. When NAD27 was first created, all measurements and calculations were done by hand and retained in hand-transcribed tables. These measurements were formulated in reference to a single ellipsoid surface datum point in Meades Ranch, Kansas. These outdated methods produced numerous inconsistencies and, in some areas of the country, geographic uncertainty due to imperfect measurements.

NADCON provides correction for some of these existing uncertainties and inconsistencies in NAD27 and, as such, provides a relatively efficient method to transform between both datums. With such a direct correlation between datums and focusing on whole grid shifts rather than individual axis or point shifts, the NADCON transformation method proves to be incredibly accurate, far more accurate than the previously discussed datum transformation methods.

The High Accuracy Reference Network

GPS and other present-day advances in geodetic measurement and coordinate position accuracy have allowed for the creation of a High Accuracy Reference Network (HARN). A HARN is a statewide or regional, high-accuracy geodetic network that improves the precision of older datum control networks. The ultimate goal for developing a HARN is to produce a denser geodetic network that will present validated results rather than human estimation.

More specific, a HARN is used to improve accuracy of NAD83 in specific state-based control networks. This increased accuracy has warranted the need for transformation grids to convert from the original NAD83 measurements (implemented in 1986) to current measured values. Prior to 1992, this datum conversion method was known as a High Precision Geodetic Network (HPGN) correction. The acronym HARN has now superseded HPGN.

In order to revise the original NAD83 measurements to the level of accuracy supported by GPS, which was introduced shortly after NAD83, HARNs were implemented throughout the United States. Starting in Wisconsin in 1989, HARNS were set up in every state to remedy and readjust the lower accuracy NAD83 coordinates. In the end, NAD83 was revised and the HARN measurements produced a coordinate set with greater local accuracies throughout the United States. Incidentally, HARN surveys are becoming more popular worldwide, with HARNs appearing in Europe and Central America.

Canadian and Australian National Transformations

Popular grid transformation methods have also surfaced outside the United States, with probably the most common and extensively implemented being the

Canadian Spatial Reference System's NTv2 software. NTv2, which stands for National Transformation Version 2, is a grid shift interpolation software that transforms coordinates (nationally) between NAD27 and NAD83.

NTv2 has been available since 1992, when it replaced the former version. The NTv2 software has proved to be immensely useful in Canada, as well as in other horizontal geodetic datum networks, and has the capability for accurate modeling based upon various national control survey and municipal networks. NTv2 has remained Canada's national grid transformation method.

In November 1997, Australia's Intergovernmental Committee on Surveying and Mapping (ICSM), faced with a recently revised national datum, turned to Canada's grid transformation method. The conversion utility was adopted for Australia's national conversion from the older, less accurate Australian Geodetic Datums of 1966 and 1984 (AGD66 and AGD84) to the GPS-referenced Geocentric Datum of Australia 1994 (GDA94).

Australia implements NTv2 to handle the differences between datums as a national grid shift, with various control stations surveyed throughout Australia. NTv2 is used to provide horizontal control when converting between the two datums and enables accurate positional coordinates through simple software-driven interpolation. Since AGD66 and AGD84 are equally used, Australia utilizes two grids: one to convert AGD66 to GDA94 and another to convert AGD84 to GDA94. In the end, implementation of Canada's NTv2 method will enable Australia to have greater local accuracy nationwide, purported to be within 10 cm.

Reference

[1] Jones, A., and G. Blick, *Where in the World Are We?*, ver. 1, Wellington New Zealand: Government of South Australia, 1999.

8

The Vertical Datum

Just as a horizontal control datum requires a reference grid on which all coordinates abide, a *vertical control datum* (or *vertical datum*) demands a contiguous, consistent reference from which topographical elevation and bathymetric depth are measured. Perhaps considered a complex geodetic construct, the vertical datum is used to manipulate and define a positional height as it relates to a determined reference. How can one accurately measure heights when the Earth's surface ranges in elevation from the peak of Mount Everest to the underwater depths of the Pacific Ocean's Marianas Trench? What could constitute a valid reference for precise vertical measurements?

These and other similar questions have plagued geodesists, mathematicians, and mariners alike. The geodesists had standard ellipsoids that proved to be a general, though innately imprecise, height reference. Mathematicians had their mathematical model of the Earth that was, for height, equally imprecise. Mariners had the tides as reference and frequently, due to the ever moving and inconsistent height of the seas, found themselves stuck in sandbars or crashed upon the rocks. The early need for a consistent and accurate height reference was desperate.

Today, vertical datums provide just such a reference from which positional heights on the Earth's topographical surface can be determined with reasonable accuracy. There are two primary reference surfaces used by leading vertical datums: sea levels (tidal) and standard ellipsoids (geodetic). References for sea level–based datums vary from high and low tide levels to an average sea level. Ellipsoid-based datums generally reference the same ellipsoid model used for calculating horizontal datums. By function, GIS utilizes these same reference surfaces when defining general and specific point heights.

Mean Sea Level and the Geoid

In many day-to-day uses, elevations are referred to as height above or below sea level. For example, in the case of the highest and lowest points on Earth, the peak of Mount Everest is 9 km above sea level and the base of the Marianas Trench is approximately 11 km below sea level. The Earth's topology varies 20 km, highest point to lowest point. In the grand scheme of ellipsoids, this variance is nearly negligible (remember the dimples on the golf ball from the ellipsoid chapter). However, when referenced to sea level, the elevation swing is perceivable and affecting. We can easily conclude that height measurements referenced to the ellipsoid are less accurate than when referenced to the sea level.

Additionally, since tides are constantly changing, rising and falling, the term "sea level" is considered the average of the tide levels. In terms of datums, we call this average sea level the *mean sea level* (MSL). Figure 8.1 illustrates this relationship.

As represented in the figure, the height measurement is the distance from the control station or geodetic point to the MSL. This is generally called the *mean sea level elevation.* Known formally as the *orthometric height,* the measurement is the height above the MSL and, depending upon position, can be represented as a positive (above MSL) or negative (below MSL) measurement.

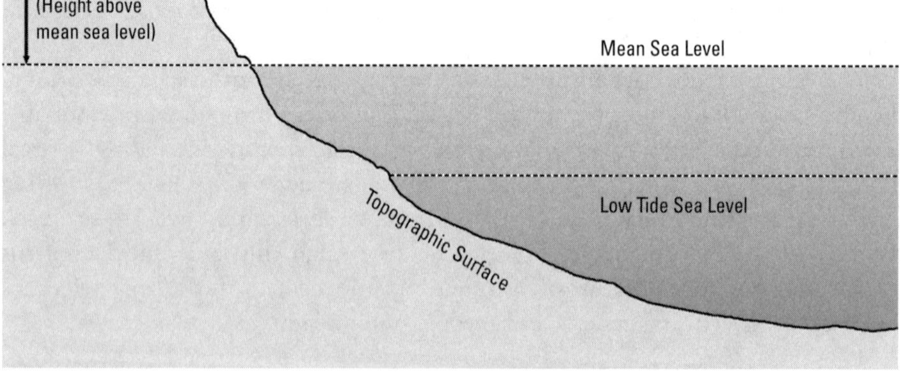

Figure 8.1 The relationship between sea level and height measurements.

As we know, tides are highly influenced by the Earth's gravity field. The Earth's gravity field is defined by two distinct components: the normal mathematical gravimetric earth model and the deviations and disturbances (known as anomalies) from this normal model. In the middle of the ocean, undisturbed by land, the MSL and the Earth's gravitational field coincide. On land, the terrestrial mass and density affect the gravity field and, thus, cause a very small variation between MSL and the Earth's gravity field. Nearly everywhere on Earth, this small variation can be ignored without diminishing the accuracy of height measurements.

Now, imagine if the MSL were to cut through the continents and create a level surface around the Earth. The result would be an exact model of the Earth defined by the MSL that would double as an adequate depiction of the Earth's gravity field. This model is known as the *geoid.*

The geoid is an equipotential (or level) surface of the Earth's gravity field, which coincides with the MSL. For all reasonable purposes, the geoid is defined by the MSL and is considered a good reference by which elevations or heights can be measured.

The geoid is not a smooth surface. The gravitational pull of the Earth is stronger in topology rich in iron and other dense materials, and, as such, causes the geoid to take on a less refined shape. Figure 8.2 depicts how the geoid looks in reference to the standard ellipsoid. Notice its irregular shape, resembling a potato.

The shape differences between the geoid and the standard ellipsoid illustrate just how dissimilar these vertical datum reference surfaces are. To understand the relationship between these reference surfaces, we must examine the separations between the accurate but complex geoid (MSL) and the general but

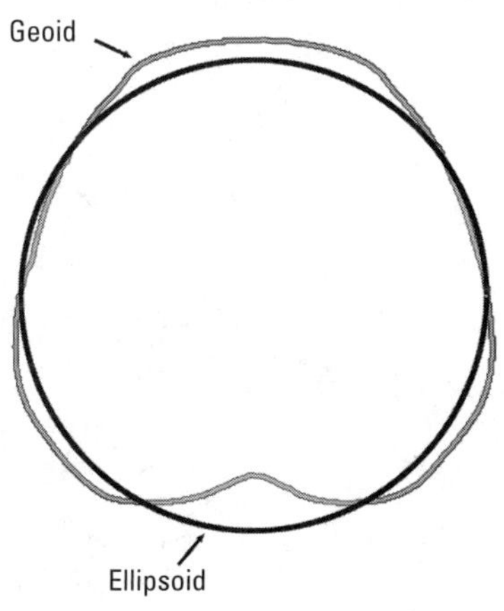

Figure 8.2 The geoid's surface.

simple ellipsoid. These surface separations are called *geoid-ellipsoid separations* or *undulations.*

The Geoid-Ellipsoid Separation

The geoid-ellipsoid separation is an important factor in defining the vertical position as it relates to the vertical datum. Since GPS and certain datums reference the ellipsoid, while other vertical datums reference the geoid, a firm understanding of their relationship is needed to link datums and determine precise vertical positioning between systems. Figure 8.3 details the specific relationship between the geoidal, ellipsoidal, and topographical surfaces

The components to vertical positioning and interrelation between datums can be determined mathematically using the following formula:

$$h = N + H$$

The geodetic height h (also called the *ellipsoid height*) is defined as the height above the ellipsoid to the topographic surface. The orthometric height H is defined as the height above the MSL. The geoid separation N is the distance between the ellipsoid and the geoid. Geoid separation (undulation) values are positive when the geoid is above the ellipsoid and negative when the geoid dips below the ellipsoid. Incidentally, throughout a large portion of the Earth, the geoid is above the ellipsoid (N is positive). In the United States, however, the geoid is below the ellipsoid and N is negative.

The compatibility between various vertical datums is paramount to controlling and manipulating the vertical component in a point position. Understanding geoid undulation enhances datum interoperability and vertical datum

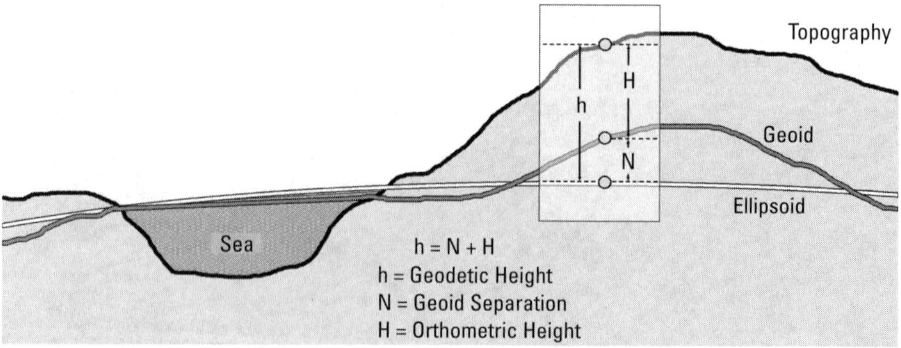

Figure 8.3 The geoid-ellipsoid-topography relationship.

transformations. Let us now examine some standard vertical datums and common transformation methods.

Standard Vertical Datums

The heights or depths of terrestrial points are determined through referencing the defined elevation surface of a vertical datum. Like the horizontal datum, the need for positional accuracy has prompted the development of standardized vertical datums. Once again, each major province or country has its own national standard datum of choice or design. They are used throughout the world for elevation-related reference and precision along the *z* axis. We will overview four commonly used vertical control datums for the Earth model, namely the Earth Gravitational Model of 1996, the GEOID03 reference family, the National Geodetic Vertical Datum of 1929, and the North American Vertical Datum of 1988.

Earth Gravitational Model of 1996

The Earth Gravitational Model of 1996 (EGM96) is the accurate collaborative product of the National Imagery and Mapping Agency (NIMA), the NASA Goddard Space Flight Center, and the Ohio State University. EGM96 is a geopotential model of the Earth, which means that orthometric height (height above MSL) is adjusted using the differential of gravity with latitude and elevation. Therefore, EGM96 offers heights that are influenced and adjusted by the Earth's gravity.

When working with models accounting the Earth's gravitation field, spherical harmonics are often used to approximate the shape and size of the geoid. A spherical harmonic is a single-valued, continuous, and complex calculation involving longitudinal and latitudinal angular coordinates. EGM96 utilizes a complete set (360 degrees) of spherical harmonic coefficients that offers accurate details of the entire geoid to regions as small as 55 km.

As a complex global reference surface, EGM96 details the variations between the MSL and the ellipsoid, discussed earlier as geoid-ellipsoid undulations (separations), and characterizes how uneven the Earth's gravitational potential is for the MSL.

Information for the development of EGM96 came from a collection of detailed surface gravity data, as well as direct satellite altimetry data, altimetry-derived gravity anomalies, and multisystem tracking of more than 20 satellites. This dynamic level of precision has earned EGM96 worldwide appeal and implementation. In fact, when EGM96 was first released, it was used as a geodetic reference to update WGS84.

As a well-defined global vertical datum, EGM96 is used as a modern reference for numerous applications worldwide, such as bathymetrical and geophysical studies.

GEOID03 and Predecessors

The U.S. Hybrid Geoid Model for 2003 (GEOID03) is a global vertical datum that calculates geodetic heights (height above the ellipsoid to the topographic surface) above the GRS80 ellipsoid. Developed by the National Geodetic Survey, GEOID03 is the latest revision to the U.S. GEOID model, superseding predecessors GEOID99, GEOID96, GEOID93, and GEOID90. With each revision came additional refinement and enhanced precision to the model. Although greatly used in the United States, GEOID03 is an excellent gravimetric geoid worldwide.

The GEOID03 model is considered a "hybrid" geoid, because it was developed using both a very precise gravimetric geoid (U.S. Gravimetric Geoid Model of 2003, or USGG2003) and updated NAD83 GPS ellipsoidal heights on leveled benchmarks (GPSBM2003). Incidentally, through this background infrastructure, GEOID03 is a derivative of EGM96.

GEOID03 remains the leading geoid model in the United States and the foundation for national models elsewhere in the world. The GEOID03 model is accurate to within 1.0 cm throughout the conterminous United States (CONUS). Since the geoid model is calculated using GRS80, it holds positional accuracy equivalent to GPS. In addition, conversion from NAD83 ellipsoidal heights to the North American standard vertical datum, NAVD88 (discussed later), are accurate to within a few centimeters.

The model's revision in 2003 is less obvious with respect to its predecessor GEOID99. Both are extremely similar, having backing from the national gravimetric geoid and NAD83 GPS benchmarks. The distinction is the countless, more explicit local definitions of orthometric heights throughout CONUS. Additionally, GEOID03 was modified in 2005 to reflect the orthometric height changes in the coastal Louisiana region resulting from a series of damaging hurricanes.

National Geodetic Vertical Datum of 1929

The National Geodetic Vertical Datum of 1929, known as NGVD29, is a geodetic reference for elevations based on a mean sea level datum. Formally known as the Sea Level Datum of 1929, NGVD29 has proved adequate for all areas in North America. Although it is an older vertical control datum that is slowly being superseded, it is still very much used throughout the United States and Canada. Many states and commonwealths still use maps based on NGVD29.

The main obstacle with using NGVD29 is that it is somewhat flawed. NGVD29 and its predecessor erroneously set a fixed elevation (zero elevation) by the MSL at 26 tide gauges in North America (21 in the United States and 5 in Canada). Since the Sea Level Datum of 1929 was first calculated more than eight decades ago, the MSL and the sea topological surface of the Earth have changed.

By May 1973, this gradual change in sea level was apparent and, technically, the implication of the old datum name was incorrect. The datum was no longer viewed as a reference to the MSL or geoid. This prompted the National Geologic Survey to change the name to the current National Geodetic Vertical Datum of 1929. Only the name changed; the outdated measurements remained unrevised. Needless to say, elevations measured using NGVD29 and Sea Level Datum of 1929 are different at different times.

During its lifetime, NGVD29 has grown flawed and has lost its usefulness in relation to positional accuracy. Geodesists can now measure these elevation differences more precisely with the latest advances in technology and a sophisticated geodetic network. Many U.S. state agencies and Canadian provinces have either begun or completed the chore of switching from NGVD29 to the more accurate North American Vertical Datum of 1988. In years to come, NGVD29 may be used less and eventually become obsolete.

North American Vertical Datum of 1988

The North American Vertical Datum of 1988 (NAVD88) is a vertical control datum first established in 1991. NAVD88 is considered a modern control datum that institutes what is called a minimum-constraint adjustment of leveling observations in Canada, the United States, and Mexico. NAVD88's minimum-constraint adjustment is a nominal modification to elevations based upon one tidal benchmark at Father Point/Rimouski in Quebec, Canada. The result is a very precise vertical control datum that is responsive with respect to geologic changes and is no longer tied to the time-altering MSL.

As mentioned previously, NAVD88 is superseding the older, less precise NGVD29. This process is not as easily implemented as one would expect given that each location in North America has a different conversion factor between the two datums. U.S. states and Canadian provinces have begun redrawing regional and height-sensitive maps (such as flood maps) with the new vertical control datum. Many regions provide area-specific conversions for use with the existing outdated maps. For instance, Wilkes County in North Carolina uses a datum offset of –0.48 feet and therefore calculates the NAVD88 elevation as the NGVD29 elevation – 0.48 feet.

Eventually, NAVD88 may be the only North American standard vertical datum used for vertical control and height accuracy. As we will discuss in the next section, transformation software has been developed to help transform height data between NGVD29 and NAVD88.

Vertical Datum Transformation Methods

As with horizontal datums, there is a functional need to transform between different vertical control datums. As we know, heights are dependent upon the reference point (or zero elevation). There are many instances when height (elevation) transformation is warranted. You might want to reference a horizontal datum height to a gravimetric geoid or you might need to transform a height from an older vertical datum to a newer, GPS-precise datum.

Many horizontal datums provide a height quantity or distance along the *z* axis. To transform heights from one ellipsoid-based datum to a second ellipsoid-based datum, the Molodensky and Bursa-Wolf transformations continue to prove adequate. Both transformation methods were discussed in the previous chapter (horizontal datums), but allow for a position point's *z*-axis (height) transformation. The level of accuracy is based upon the ellipsoid origin and *z*-axis datum shift. This type of vertical position transformation does not account for the Earth's gravity field or the geoid.

A transformation from a horizontal datum height to a vertical datum height is sometimes required. These transformations involve referencing a gravimetric geoid, like GEOID03, between the horizontal and vertical control datum heights. For instance, transforming vertical positions from NAD83 to and from NAVD88 requires the typical height transformation with an implementation of GEOID96 or later. The transformation converts the position and the geoid model factors in the gravity field's potential.

The last form of vertical height conversion is a true vertical datum transformation—transformation from one vertical datum to another. The general impetus for implementing such a conversion is to transform older datum positions to newer, more precise positions.

The most common implementation of this transformation is between NGVD29 and NAVD88. To convert easily between the two vertical datums, the National Geodetic Survey devised VERTCON, a vertical model transformation method that computes the difference in orthometric height between NGVD29 and NAVD88. VERTCON was implemented to help state and county agencies revise their maps and charts from the older datum to the newer, GPS-based datum. Both datums utilize the same coordinate grid; therefore, VERTCON is considered a vertical grid transformation model.

The VERTCON model calculates the *z* value at any point in CONUS through the following calculation: $z = \text{NAVD88} - \text{NGVD29}$. VERTCON highlights, through this simple mathematical equation, the varying inaccuracies in the older NGVD29 datum. The VERTCON method utilizes a preprogrammed grid on which a transformation is calculated at every point on the ellipsoid. The resultant z values serve as a reference for every location in CONUS.

If you have an NGVD29 height to convert, the VERTCON model z value for that location should be added to the NGVD29 height (i.e., NAVD88 = NGVD29 + z). Alternately, if you have an NAVD88 height to convert, the VERTCON z value for that location should be subtracted from the NAVD88 height (i.e., NGVD29 = NAVD88 – z).

The VERTCON method is a common, much-used vertical datum transformation model that has warranted its own software. Whatever method you employ for height transformation, the end product will most likely be a more defined height position that not only will conform to modern accuracy standards but also will depict the precise orthometric heights in reference to a common Earth model.

9

The Map Projection

When was the last time you went into a gas station and purchased a projection of the Earth? Believe it or not, it was the last time you bought a road map. Atlases, world maps, and road maps are all projections of the Earth under well defined criteria.

The idea of depicting the Earth on a flat map is easy to imagine and rather simple to understand, but the actual process of a map projection is complex. Maps used in everyday life, as well as maps used in GISs, are created with a specific application in mind.

Mapmakers, or *cartographers*, use map projections to portray the real world with minimal aberrations. This chapter specifies these tools in detail and defines projection techniques and understandings that are not readily discernible from an everyday road map.

The Physicality of the Earth

Before discussing the *what* and *why* of a map projection, let us first recap a fundamental element of Earth science: the ellipsoid. The Earth's ellipsoid rotates upon a single, standard axis, making one full revolution every *sidereal day* (or 23 hours, 56 minutes, and 4.09 seconds). The Earth's overall shape is commonly defined and modeled as an *ellipsoid of revolution.*

More specifically, this axial rotation causes every location on Earth to experience sunlight and darkness, although within the artic and antarctic circles, the axial rotation creates increased lengths of sunlight and darkness. The Sun faces only a portion of the Earth at any given moment. This illuminated portion, over time, rotates beyond the range of sunlight and enters darkness. Almost

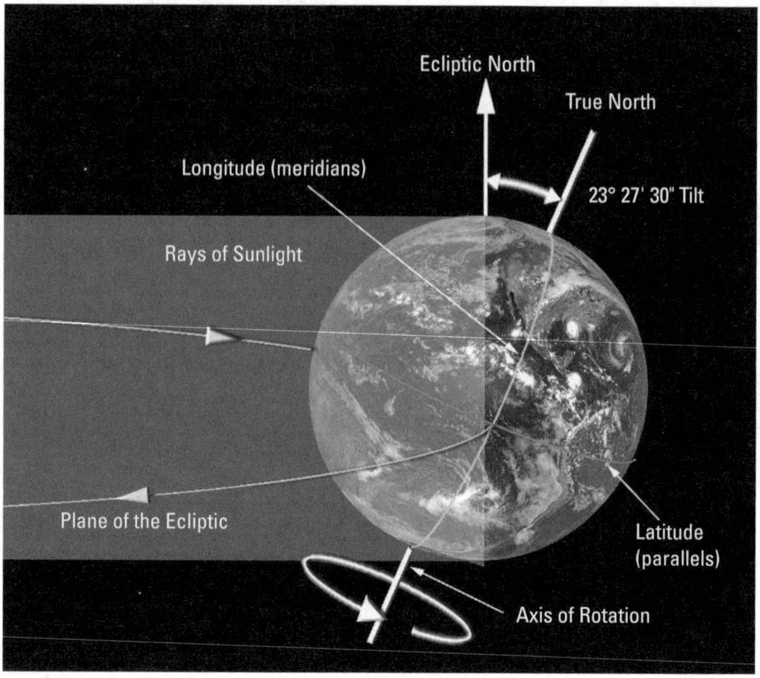

Figure 9.1 The physical Earth.

all locations on Earth experience sunlight and darkness in one day, but areas close to the poles (within the arctic and antarctic circles) experience longer daylight and nocturnal phases. This variance is a result of the Earth's tilt, which extends or diminishes a location's exposure to the Sun (see Figure 9.1).

The standard axis is perpendicular to the Earth's surface and intercepts the ellipsoid's surface through two distinct points known as the Earth's poles: the north pole in the arctic and the south pole in the antarctic. Figure 9.1 indicates this "axis of rotation."

The Earth revolves around the Sun with a single revolution taking 365.2564 days (known as a *sidereal year*) and defining a plane in space called the *plane of the ecliptic*.

The Earth's standard axis runs north and south and is tilted 23 degrees 27 minutes 30 seconds from the plane of the ecliptic. This tilt is the reason we experience the four seasons of the year. For example, in the summer, which starts with the summer solstice (June 21), the northern hemisphere is tilted toward the Sun, causing a rise in temperature. During the winter, which starts with the winter solstice (December 21), the northern hemisphere is tilted away from the Sun, causing a fall in temperature.

The Earth's ellipsoid is partitioned by a network of horizontal (east/west) and vertical (north/south) lines. The horizontal lines are called *parallels,* because they run parallel to one another, while the vertical lines are called *meridians.* The Earth has a prime meridian, which runs through Greenwich, England, and a central parallel called the equator. The equator separates the northern hemisphere from the southern hemisphere. In terms of the Earth's ellipsoid, each location along a parallel has the same *latitude,* or distance from the equator. Similarly, each location along a meridian has the same *longitude,* or angle between the target meridian and the prime meridian (0°). Although the meridians converge at the two poles, the parallels never converge.

The Earth is a complex three-dimensional object with physical dimensions, including height, width, depth, mass, and density. An important physical characteristic is its *curvilinear surface,* which is a surface consisting of or bounded by curved lines. The Earth's nonstraight, curvilinear surface is the sole reason why a latitude and longitude system is necessary. Figure 9.2 depicts the Earth model with its geographic coordinate system.

This curvilinearity also adds complexity to a map projection's transformation of the Earth model or *globe* to a flat surface. In fact, it is the primary reason why map projections are only cultured depictions of the globe and not exact representations. Due to the Earth's curvilinear surface, map projections always hold a degree of distortion.

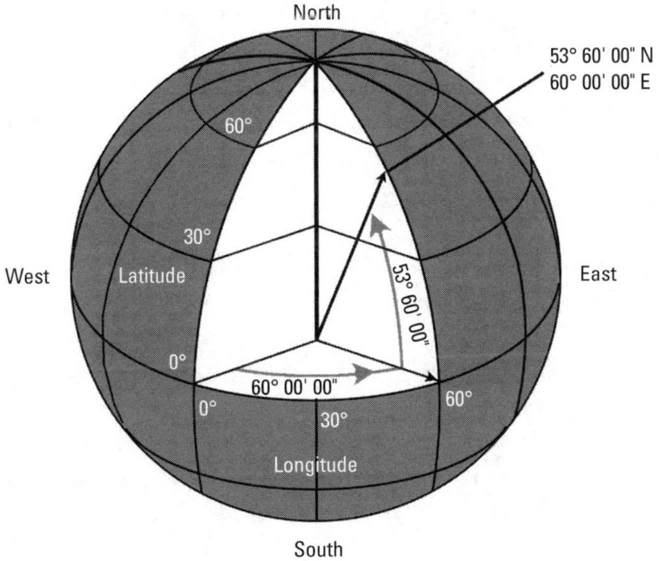

Figure 9.2 Earth model geographic coordinate system.

Scale Reduction: From the Ellipsoid to the Globe

Up until now we have discussed the Earth's ellipsoid in relation to its physicality and true size, but for the purposes of geospatial analysis the Earth's true dimensions are unwieldy. They are simply too large to realistically handle, manipulate, and compute. All GIS applications ultimately use an interpretation of the Earth at a reduced scale. For this reason, before a map projection can be created, the dimensions of the Earth must be reduced.

The process of proportionately reducing actual dimensions into much smaller, more convenient dimensions is called *scale reduction.*

Scale reduction depicts the projecting ellipsoid as a drastically scaled-down model, much like a tabletop globe of the Earth. The Earth's globe offers the user an easily accessible object, capable of depicting a range of geographic data at flexible vantages. Relative to GIS applications, the Earth's globe is typically used in the creation of map projections, since the ellipsoid proves unmanageable.

Let us look at the use of globes in map projections in a more fundamental way: Think of the Earth's globe as being made of transparent glass with a light emanating from its center in a darkened room. The images appearing on the walls in this room are projections of the globe, similar to the shadows around a lampshade as its light bulb burns bright. This Earth model is easily manipulated and projected and properly characterizes the importance of and need for scale reduction.

A scaled-down Earth is absolutely necessary for any GIS application. For instance, a road map printed on a 50-foot mainsail will do you very little good in the car at the stoplight. A letter-size map of the United States will prove futile in assisting you in finding the downtown of a major city from the suburbs. In summary, scale reduction must be used in the creation of map projections to transform the bulky ellipsoid of revolution into the more manageable and easily projected globe.

Transformation from the Globe

The basic purpose of a map projection is to transform the Earth's globe into a useable format for a specific application. A projection systematically transforms the geographic latitudes and longitudes on the surface of the ellipsoid into locations on a plane. Locations in three-dimensional space are arranged to correspond to a two-dimensional representation. Think of the Earth's globe in relation to a flat map of the world.

Projections that perform these systematic transformations from the globe are classified as *geometric.* Geometric projections are subclassified and defined according to the type of object the projection uses.

The geometric map projection methodically transforms all geographic coordinates on a globe into corresponding plane coordinates of a map space. Figure 9.3 illustrates this transformation. A basic projection formula can be expressed as

$$X = Function\,1\,(Latitude, Longitude)$$

$$Y = Function\,2\,(Latitude, Longitude)$$

X and Y positions on the projected surface are functions of latitude and longitude, commonly represented by the Greek letters lambda and phi (λ and Φ), respectively. These functions must be unique to ensure that a particular point will appear at only one position on the map. They must be finite, or within a limit, so that a particular point will not appear at infinity. They must also be continuous, ensuring that although stretching or shrinking may occur, no gaps are introduced.

The geometric projection uses this formula and these projection methods to initiate a transformation, transferring latitudes and longitudes to locations on a surrounding surface.

What Is a Map Projection?

A *map projection* portrays a three-dimensional object, such as the Earth's globe, in a two-dimensional format. The map projection is quite simply the most intriguing component of coordinate system referencing because it offers a high

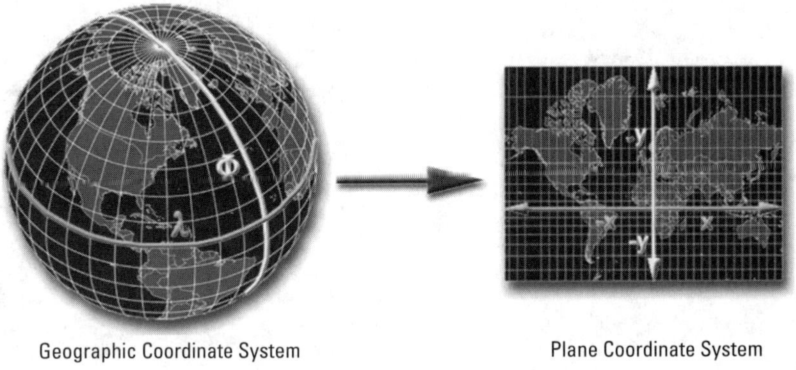

Figure 9.3 Moving from the Earth's geographic coordinate system to the projection's plane coordinate system.

level of flexibility. Projections are wholly graphical, as evident in the highly used Goode Homolosine map projection of the world (Figure 9.4).

The map projection employs projection formulas that perform the critical task of transferring a three-dimensional spheroid onto a two-dimensional plane surface. This is a complex process because the Earth, like other celestial bodies, is a complex object. While the projection is the easiest coordinate system component to visualize, the projection application is the most difficult. As you will quickly discover, an understanding of the projection application is vital to the success and usability of the transformation.

Map projections, by default, are not true or accurate portrayals of the globe. A two-dimensional plane cannot accurately represent large portions of the rounded, curvilinear surface of the Earth. Figures 9.5 and 9.6 illustrate the Mercator model and projection. Notice the shapes of North America and how different they are from in each other.

To show regions of the Earth on any appreciable area with accuracy, geographic data must be drawn to compromise the distortions of shapes, distances, and directions introduced by the spheroid. The various methods of preparing a two-dimensional plane from the surface of the Earth are critical for the accessibility and presentation of GISs.

The map projection is an important element in GIS. As discussed in the previous two chapters, we can comfortably surmise that datum superimposed upon the surface of the Earth's globe, or *ellipsoid of revolution*, establishes vertical and horizontal control for the specific area. The projection transforms this specific area from the curvilinear surface of the ellipse to a flat plane upon which an image is projected. This projected area is then implemented within a GIS.

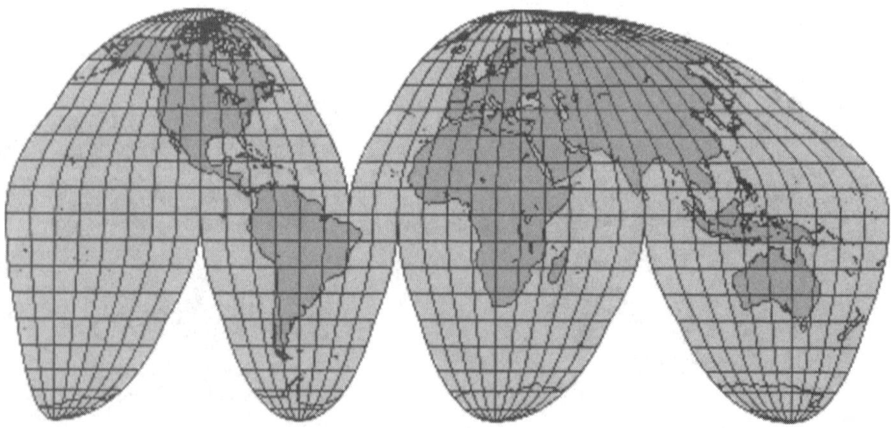

Figure 9.4 A Goode Homolosine 10° equal-area map projection of the Earth.

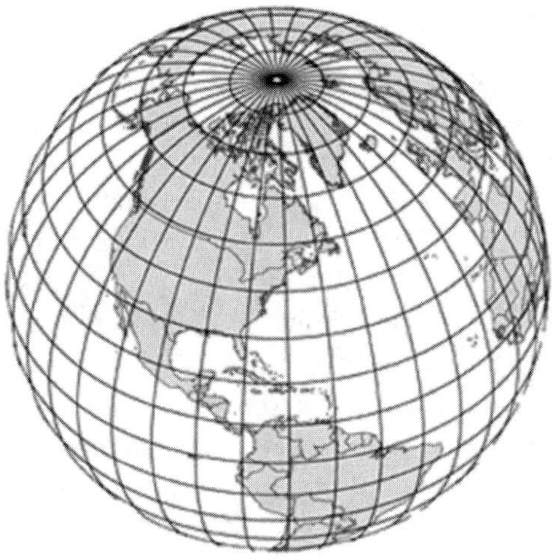

Figure 9.5 Geographic Mercator model.

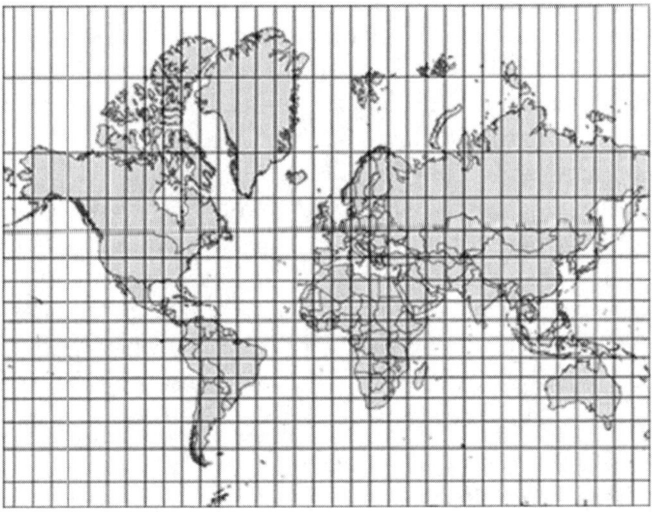

Figure 9.6 Planar Mercator projection.

The Grapefruit Peel Experiment

Let's look at the map projection a different way. Take an ordinary grapefruit and observe its spherical form. This tasty spheroid is a three-dimensional object that is similar, in form, to the Earth's globe. We will make a physical projection of one half of the fruit to help visualize a GIS map projection.

Using a flexible measuring tape and marker, draw a two-inch vertical line on the peel close to its "pole." With a knife, cut the peel perpendicular to the drawn line in one straight cut all around the grapefruit. The drawn line should be away from the peel cut. You can easily imagine each half of the cut grapefruit as a hemisphere. Now, carefully lift off the peel hemisphere (the one with the drawn line) and place it on a flat table so that the widest area is face down on the table. Notice that this object is still three dimensional and has retained the approximate curvature of the grapefruit.

Using the knife, cut the grapefruit's skin from the pole to the edge, keeping away from the drawn line. This step helps control where the skin will rip. With your palm, push down on the peel until it flattens. Creases and tears will form in various locations to compensate for the flattening of the curvature and to conform to the table surface. Once completely flat and neglecting the peel's thickness, the object has theoretically become a two-dimensional representation of the grapefruit hemisphere. Now let's examine this physical projection.

The flattened grapefruit skin does not look like a spheroid anymore and, with the peel's control cut and tears, has become distorted. Take that measuring tape and remeasure that two-inch drawn line. You will notice it is slightly less than two inches now, or that it is torn. The flattening process distorted the line's length, which is a measurement of distance, one of the four major map distortions discussed latter in this chapter.

You have now made a physical (and conic type) projection of the grapefruit. A GIS map projection functions in much the same way as the grapefruit peel experiment, though not quite as crude. When projecting a three-dimensional object onto a two-dimensional plane, various distortions and aberrations occur to functionally portray the object. Maps of the Earth's globe do not precisely depict the real shape of each continent but project the adjusted flat version of the curved landforms. Unlike the grapefruit, tears, aberrations, and distortions can be easily manipulated through projection equations.

Choosing a Map Projection

One of the more difficult parts of the map projection process is the selection of the best projection type for the application. Not only must you fully understand the object being transformed onto the flat surface, but you must also understand the desired properties you want to exhibit.

When choosing a projection, the purpose of the application must first be recognized. Then, once the purpose is understood, a strategic plan can be developed to determine those features that need to be preserved and those features that can be compromised by relative distortion. Even variations of the same projection can be selected for more accurate depictions in specific applications, such as the two Bonne projections in Figure 9.7.

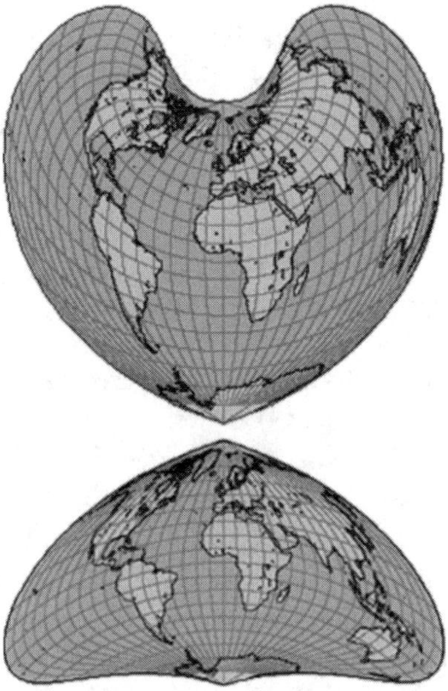

Figure 9.7 Two variations of the Bonne projection, a large-scale topographic map projection: central parallel 45° north (top) and 15° south (bottom).

The projection process is, by far, not a natural rearrangement of aspects from the three-dimensional surface. It is a mathematical condition superimposed on a natural surface to accommodate a man-made interpretation. As a result, the projected plane always has some degree of inherent distortion.

There are four important properties of the Earth model that can be tactically preserved during a projection process: *area*, *shape*, *distance*, and *angles*. These four elements are the essential pawns when projecting a surface. You must come to terms with the hard truth involved in portraying the spherical shape of the Earth on a nonspherical surface: Preserving accurate representations of all four elements simultaneously is impossible.

Important, even vital, characteristics for any projection must be identified and specifically defined early in the process. The proper type and classification of projection capable of preserving these characteristics should be tested, selected, and then implemented. If time is taken to accurately plan a projection, control of the projection will be retained and the various ill side effects from the process will be minimized.

Choosing a map projection involves more than planning; it involves decisions and an understanding of how best to represent the spheroid. Aside from manipulating the four projection elements, there are other choices and degrees of flexibility available when creating the best map projection for a particular application. The ability to choose the projection's *type*, *aspect*, and *classification* is the ultimate tool for achieving the desired projection properties and for controlling distortion. The following discussions categorically detail each decision element.

Types of Map Projections

As tools in the projection toolbox, there are three distinct types of projections that can be utilized on the projected plane: *cylindric*, *conic*, and *planar*. Commonly known as *developable projection surfaces*, these types are selected on the basis of characteristic curves. The projected plane inherits the display characteristics associated with each projection type and forms what is called a *graticule*. A graticule is basically a grid of meridians and parallels.

Each projection type implements a particular graticule as a base projection surface. During a projection, a three-dimensional object is transformed onto the projection type graticule and takes advantage of the built-in characteristics. Let us discuss each major projection type further.

The Cylindric Type

The *cylindric* projection type takes on the appearance of a rectangular graph with an x (parallel) and a y (meridian) axis (see Figure 9.8). The globe's longitudal meridians and latitudes are represented by equidistant, parallel straight lines that intersect one another at right angles. The cylindric type projection is a strict grid representation of the curvilinear surface that is true at the equator and increases in distortion toward the poles.

The Mercator projection is a good example of the cylindric type. The graticule consists of equally spaced meridians but unequally spaced parallels. As the parallels get closer to the poles, the spacing between each becomes wider. With this increase in width, there is an increase in distortion. Therefore, the distortion increases as the projection moves toward the poles. This projection type also safeguards the integrity of features at the central parallel (such as the equator in a world projection) and the regions in proximity.

The Conic Type

The *conic* projection type is fan shaped, characterized by an upside down cone over the sphere. The meridians are represented by a system of equally inclined concurrent straight lines, and the parallels are represented by two concentric circular arcs. The angle between any two meridians on the projection is less than their true difference in longitude on the sphere. Refer to Figure 9.9.

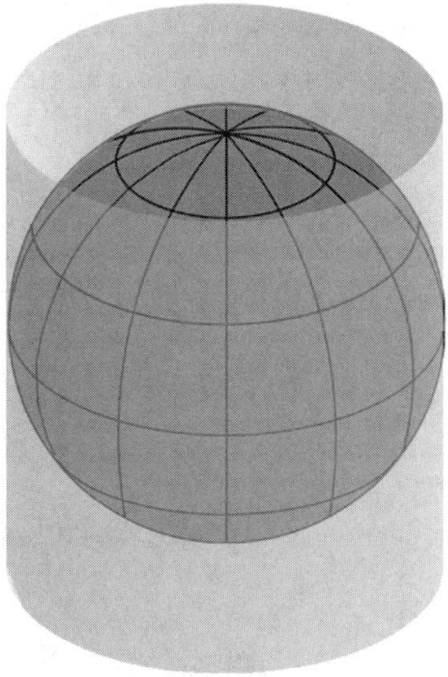

Figure 9.8 The cylindric projection.

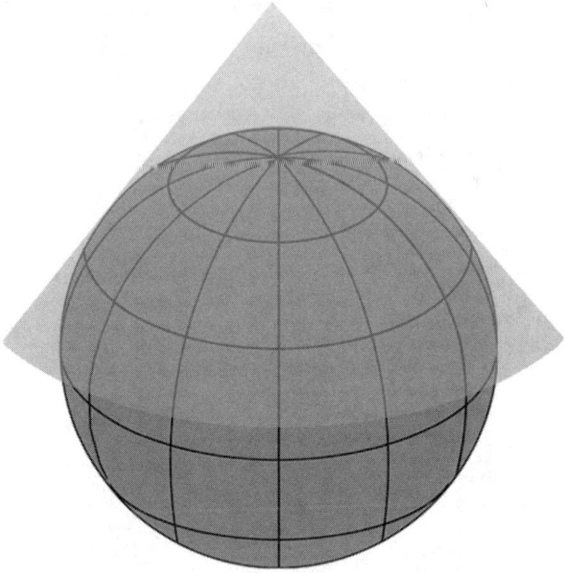

Figure 9.9 The conic projection.

Conic classifications are at an exact scale along a particular parallel, or standard parallel, between the equator and a pole. Distortion increases away from the standard parallel. According to famed map projectionist John P. Snyder, a typical conic projection will require the definition of two true-scale standard parallels demarking an area of interest.

GIS conic projections will match between geographic data sets only if the standard parallels are the same for every geographic data layer. Current popular conic type projections are the Lambert conformal conic, also known as the American polyconic, used in state plane systems, and Albers equal area, used in small-scale mapping.

Planar Type

The *planar* projection type is an azimuthal orientation of a projected surface, which means that it pertains to the angle (usually in degrees) of an object around the horizon that is measured from north to east. The planar type protects the integrity of azimuths, bearings, and directions from a central point to other more remote points in the plane. Therefore, planar projections are true only at their center point.

The planar graticule represents meridians as straight lines that incline toward each other at their true longitudal difference. Parallels are represented by a system of concentric circles with their common center at the point of converging meridians or pole. The distortions in the planar type tend to be most prevalent along the edges of the projected plane surface. Figure 9.10 illustrates a simple planar projection scheme.

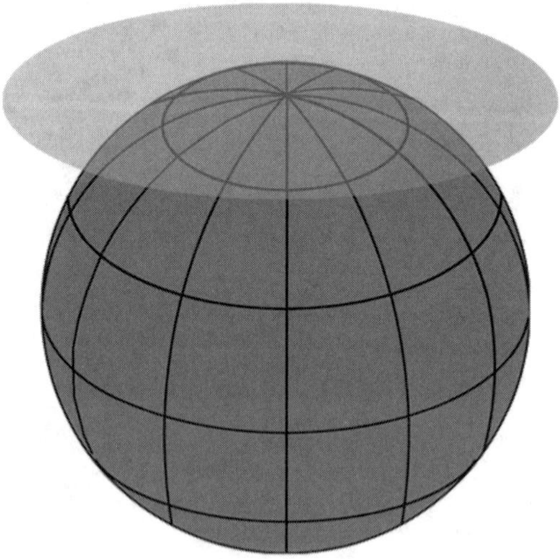

Figure 9.10 The planar projection.

The Aspect of Projection

Projection *aspect* is the relative orientation of the projected plane and the ellipsoid of revolution with respect to the location of an observer. Aspect is a natural projection property. Common with every projected surface is a standard aspect—a point or line of bearing and magnitude on which any given projection is centered.

Take another grapefruit and draw a large dot to represent a pole on opposite ends. First hold it out straight, observing its side with the poles to the north and south. Now twist it around and observe different views of the spheroid. You have the ability to observe the grapefruit from any side, from either pole, at any angle, and somewhere in between. This is exactly what a projection aspect is like; you can view the map projection from various viewpoints by manipulating the angle.

Bringing this projection tool into perspective, there are four predominant types of projection aspects: *normal*, *transverse*, *oblique*, and *polar*.

The *normal aspect* (Figure 9.11) is based upon a standard line along which distortion is minimized. The farther you move away from the standard line, the more the distortion. If this standard line is coincident with the equator, or central parallel, its aspect is considered to be *equatorial*. A normal aspect is considered *azimuthal* if the standard line connects to one of the poles while the axis of

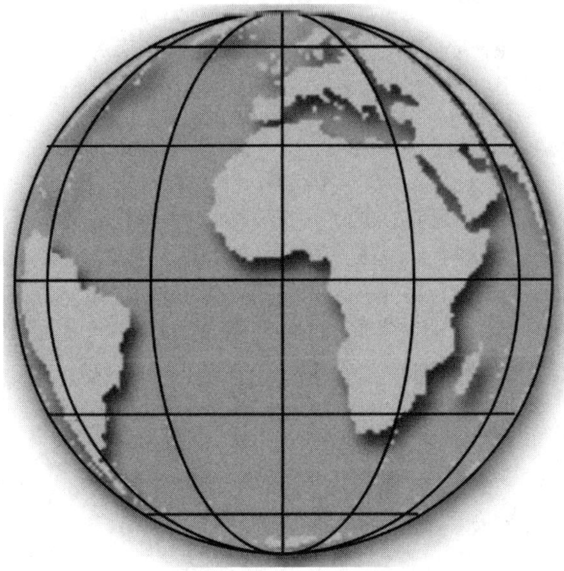

Figure 9.11 Normal aspect.

the projection (conic or cylindric) is coincident to the axis of the ellipsoid of revolution.

An aspect is said to be *transverse* (Figure 9.12) when a standard line (normal) aspect is rotated 90 degrees. This transverse aspect projection minimizes distortion along meridians and can have azimuthal or equatorial alignment.

In the *oblique* aspect (Figure 9.13), the axis of the Earth and the axis of the projection are oriented in an arbitrary manner. The oblique aspect is uniquely suited to be used for geographic areas that are centered along lines that are neither meridians nor parallels but are assumed to be "great circles" passing through the region. Distortion is minimized along the central or prime meridian in a projection. An oblique aspect is often difficult to represent as the graticule for the projected plane. Of the nine potential directions, oblique aspect types tend to be directed northwest, northeast, southwest, and southeast.

The *polar* aspect (Figure 9.14) centers on a single point, either the north pole or the south pole. Meridians are subsequently depicted as straight lines radiating from this pole outward toward the equator. Angles between these meridians are always true in relation to the longitudal distance on the Earth's globe. Latitudes (parallels) are equally spaced concentric circles radiating from this same source pole. Polar aspect projections retain symmetry about any meridian and distort between them.

Classification of Map Projections

The classification of map projections, or *projection classifications*, is used to further organize projection types and serve as another tool to gain better results for

Figure 9.12 Transverse aspect.

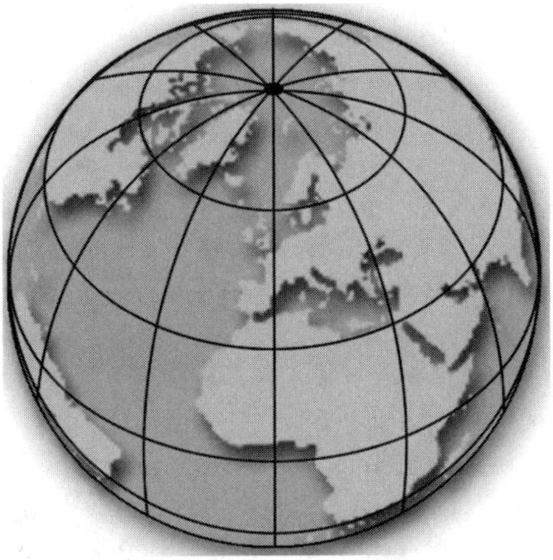

Figure 9.13 Oblique aspect.

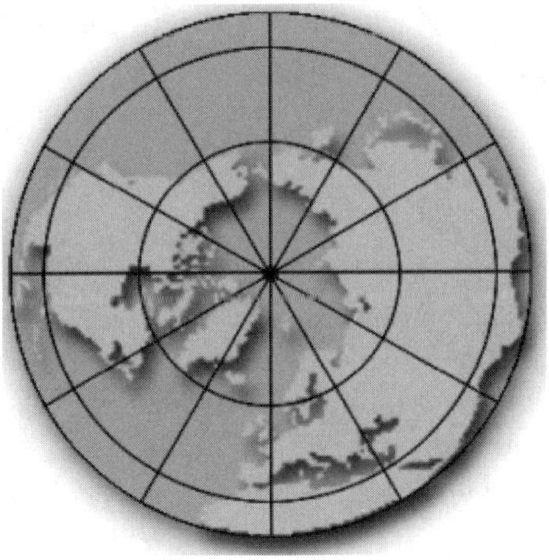

Figure 9.14 Polar aspect.

an application. While the *types* determine the specific method of physical projection, the *classification* defines the properties of the projected surface. Projection classifications ensure that not only that the desired traits required of the application are satisfied but also that the desired parameters are preserved.

There are four primary projection classifications:

1. Equidistant;
2. Azimuthal;

3. Equal area;
4. Conformal.

Projected surfaces can exhibit the properties of several classifications simultaneously, though any surface will potentially distort shapes, areas, angles, or distances at some level. Since all distortions can be measured or estimated, the selection of an adequate projection requires specific knowledge of the application. As a result, the three types of projections (cylindric, conic, and planar) can be combined with one or more of the projection classifications to control the appearance and distortion for any particular application.

For example, a planar equal-area projection preserves both angles and areas. The region of interest is accurately represented in size while angles to other locations are also preserved. The *distances* between the area of interest and other locations are absorbed by the distortion.

To wholly understand how these classifications aid in accurately projecting three-dimensional objects, each is discussed in detail.

Equidistant Projections

A projection is considered *equidistant* (Figure 9.15) when scale is true along at least one line (a focal line) or from one or two points (focal points) to all other points on the projected surface. The focal points are usually at the projection center or some other approximately central location on the surface.

Although no projection can provide a developable surface with a perfectly uniform scale, equidistant projections retain a true scale along these explicit focal features (focal lines and points). Consequently, if these focal features are properly developed, the entire projection surface will benefit from their true scale. For example, a sinusoidal projection can be classified as equidistant because all parallels and the central meridian are shown at true scale.

As another example, an equidistant projection centered on Bangkok, Thailand, will provide true distances from Bangkok to any other city in the world, such as Dallas, Texas, or Dublin, Ireland. However, that same projection will distort the relationship of distance between Dallas and Dublin.

Azimuthal Projections

In an *azimuthal* projection, all points in relation to a central point (such as a pole) are not deformed during the projection process from globe to plane. The central point occurs at the intersection of the tangential plane to the ellipsoid of revolution, or, simply, where the projection plane and the projected object meet. Azimuthal projections preserve the angular relationship of all features in a plane to the central point. This central point has zero distortion because it is the

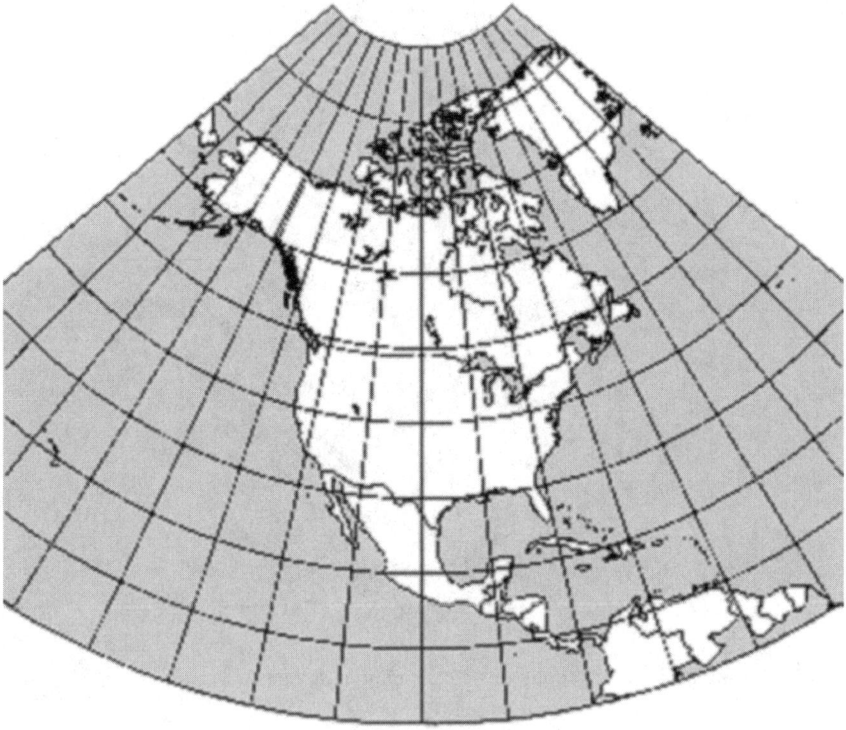

Figure 9.15 An equidistant conic projection of North America.

only point that is being truly represented in the projection. Consequently, distortion gradually increases as distance from this central point increases.

Azimuthal projection classifications are somewhat limited in that they correctly depict directions or angles to all other points in the projection with respect to one or rarely two central points. Figure 9.16 has only one central point, the North Pole.

Equal-Area Projections

The classification is considered *equal area* when the relative sizes of all features on a globe are maintained during the projection process. An equal-area projection vigilantly retains area properties of the spheroid on the plane surface through the use of compensatory scale factors. This means that if area is to be preserved but scale cannot be, then any given feature on a globe (such as a state) requires a scale factor greater than 1.0 in one direction and less than 1.0 in the other direction. Both scale factors compensate for one another and, in doing so, retain the area characteristic.

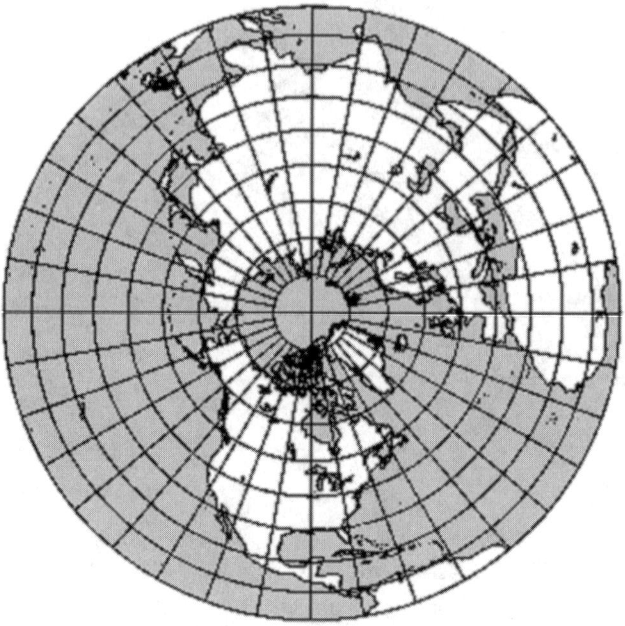

Figure 9.16 Lambert azimuthal equal-area projection of the Earth.

Using compensatory scale factors in equal-area projections, a circle on a globe will project as an ellipse on a plane surface. For example if a coin is placed on any area of a projected surface, it will cover as much of the surface of the sphere as if it were placed elsewhere on the projected surface.

When an equal-area projection is used for small-scale applications that show larger regions, such as in Figure 9.17, the distortion of both angles and shapes increases as the distance of an area from the projection origin increases. Although multiple projection classifications can be concurrently employed, a projection surface cannot exhibit equal-area and conformal classification properties simultaneously.

Conformal Projections

Unlike equal-area projections, *conformal projections* have equal scale factors in all directions at any one point on the projection surface. Instead of preserving area or size, conformal projections preserve shape. For this reason they are also referred to as *orthomorphic* projections.

A conformal surface is also defined as a plane on which all angles at infinitely small locations are correctly depicted. This projection classification increasingly distorts areas away from the point or lines of true scale (a scale

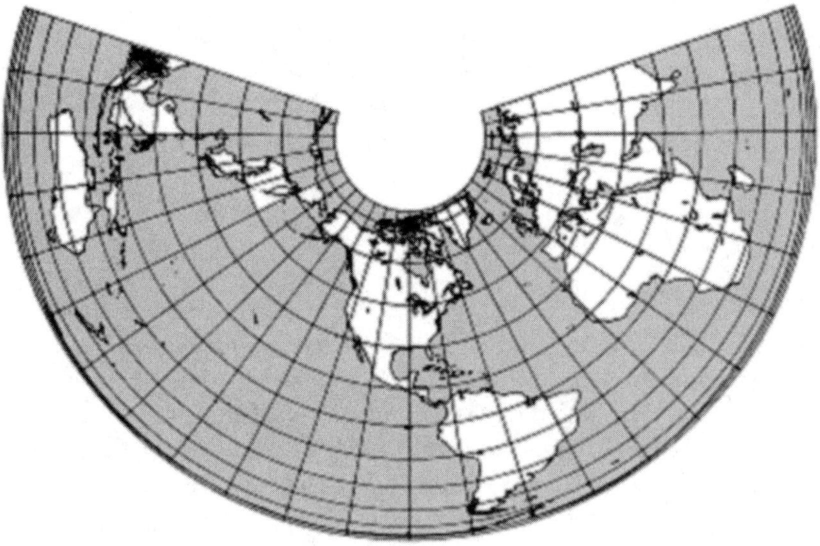

Figure 9.17 Albers equal-area conic projection.

factor of 1.0) on the plane. A good example is the Lambert conformal projection (Figure 9.18). Conformal projections, like the Lambert projection, are ideal for medium- to large-scale projection applications.

Conformal projections are also ideal for constructing plane coordinate systems for use in regional grids. These projections are conducive to managing groups of adjacent systems. Since there is a single scale factor controlling the

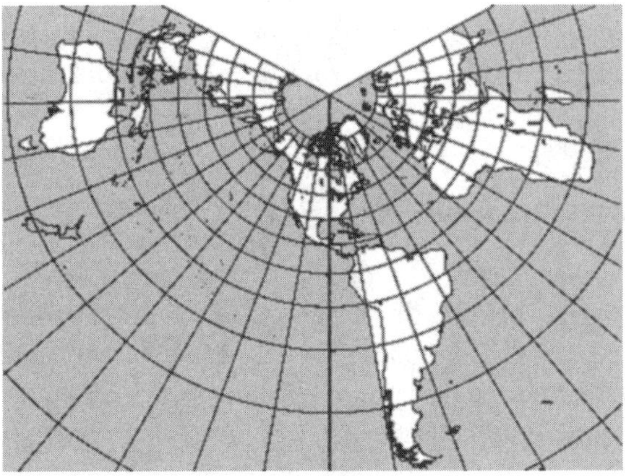

Figure 9.18 Lambert conformal conic projection.

projection, angles retain their integrity in the projected surface. This scale factor, generally not true scale, applies to all distance measurements, unless the distance is of great length or precision. For example, a circle on the globe will render as another, smaller circle in the conformal projected surface.

Distortion

Now that we have discussed the basic tools to transform a three-dimensional object and manipulate the projection, we must look at distortion. We know that the presence of feature distortion is unavoidable when a three-dimensional surface is reduced and projected to a two-dimensional plane. In order for a feature or set of features to be projected without appreciable distortion, it must be *explicitly accommodated* by the entire projection structure (i.e., classification, type, aspect, and so on). Features that are not explicitly accommodated may or may not incur the distortion of geometric properties, regardless of the intention.

The goal of any projection is to minimize the effects of unwanted distortion; therefore, a clear segregation of explicitly and nonexplicitly accommodated features must be made. Geometric integrity is compromised or sacrificed in one feature for the benefit of another. The distribution of these features varies by situation and the specific application.

Distortion Type

Projection distortion is a window into the projection, where its true strengths and weaknesses are revealed. Though most projection distortion is angular or *areal*, a projection can potentially distort five spatial properties of a developable surface: *area*, *angle*, *shape*, *distance*, and *direction*.

For each of these spatial properties there are two primary considerations: How much distortion is present and what is the effect of distortion in the projected surface?

Distortion Magnitude, Distribution, and Overall Effect

Distortion magnitude, or the amount and density of distortion in a projection, answers the question: How much distortion is present? A general rule of thumb is that the larger the area, or *region of interest*, the more pronounced the distortion becomes. This is particularly noticeable in small-scale applications, where distortion magnitude is at its peak. A fine example of a small-scale application is the world map. In small-scale applications, the projection area is large and the scale factor is small, whereas in a large-scale application the projection area is small and scale factor is large.

Small-scale distortion, especially in the world map application, is an issue that has captivated scientists for nearly 3000 years. In fact, Greek astronomer Claudius Ptolemy (circa 150 A.D.) worked with geospatial projections on a small scale, while projections, in general, are known to have been in use some three centuries earlier.

The *overall distribution* and effect of substantial geospatial distortion is often apparent to the eye. By just looking at the projection, you can see where the distortion occurs and how this distortion affects the surface features. For example, a Mercator projection of the world shows the area of Greenland larger than the area of South America, when in reality the area of Greenland is a fraction of the area of South America. These landforms are profoundly affected by the increasing distortion of the Mercator projection toward the poles. The Mercator projection retains the integrity of angles but not areas, resulting in a distributed surface that visibly distorts the appearance of substantial northerly and southerly features.

As discussed previously in this chapter, projection classifications can be used to minimize certain forms of distortion. Classifications enable the user to manage the overall distortion distribution and effect for a specific application. For instance, if a Mercator is projected on an equal-area surface, the area of landforms is retained while the other feature characteristics, such as angles, shapes, and distances, are compromised. Manipulating distortion for a particular application is your best means of properly creating and using a projection.

The Tissot Indicatrix

In 1881, cartographer Nicolas Auguste Tissot published his groundbreaking approach to distortion analysis, now known as the *Tissot indicatrix*, in *Memoire sur la Representation des Surfaces et les Projections des Cartes Geographiques.* The Tissot indicatrix is a helpful tool for visually modeling distortion and for graphically analyzing the distortion properties in a projected surface. Tissot's approach to distortion analysis was clearly the most innovative in his day and still remains an effective method of interpreting a projection. He first began describing this approach as early as 1859, and it has now joined other nineteenth-century mathematical and geographic innovations like the least-squares method or conformal mapping.

The success of the Tissot indicatrix resides in that it is a simple method for visualizing distortion. It has two primary elements: the projected graticule and geometric deformation indicators. Simply stated, the indicatrix depicts the projected graticule as a distortion graph and shows distortion as circles or ellipses at the specific intersections in the plane. A circle represents conformality (no distortion in shapes), while the relative flattening of an ellipse represents the extent and type of local distortion at the intersection where the object is plotted. These circular or elliptical objects, known as *geometric deformation indicators*, are fixed at a finite scale with explicit locations in the projected surface.

Figure 9.19 depicts a basic Tissot indicatrix with deformation indicators. Figure 9.20 illustrates a standard transverse Mercator projection with its related Tissot indicatrix. Notice how the distortion indicators become larger as they move away from the two central points.

Geometric deformation indicators are circles or ellipses that represent the distortion scale factor at the intersection of any perpendicular graticule line set on the ellipsoid of revolution. The determination of such indicators is the primary driving concept of Tissot's indicatrix and projected surface deformation theory.

Tissot also treats the projected space as a plane with the same size as the region of interest on the reference ellipsoid. For instance, an infinitely small circle on the surface of the Earth model (reference ellipsoid) is identical to an infinitely small circle on the surface of the projected plane.

For the purposes of enhanced visualization and practical appearance, the indicators are significantly enlarged in the indicatrix graph. Although not drawn to scale, the indicators are scaled proportionately to other indicators in the same solution. Accordingly, these geometric deformation indicators depict the characteristics of the projected plane at or nearly at the intersection where they are plotted.

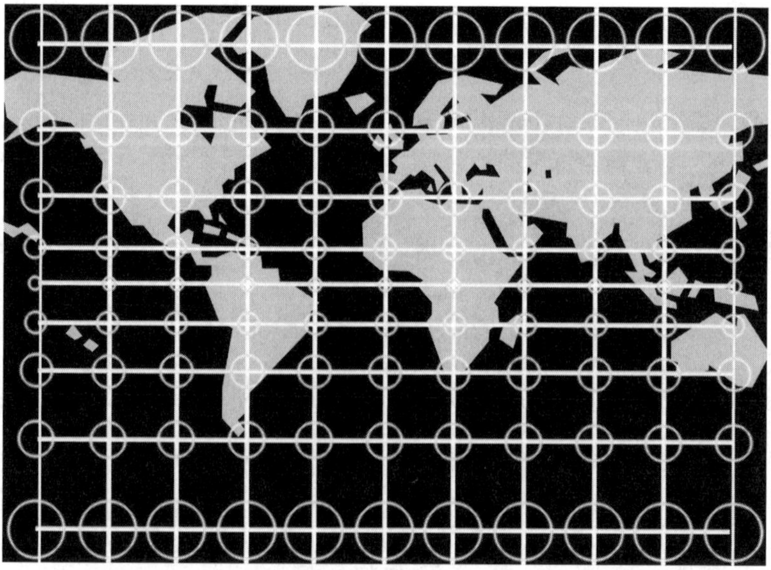

Figure 9.19 A simple Tissot indicatrix of the world.

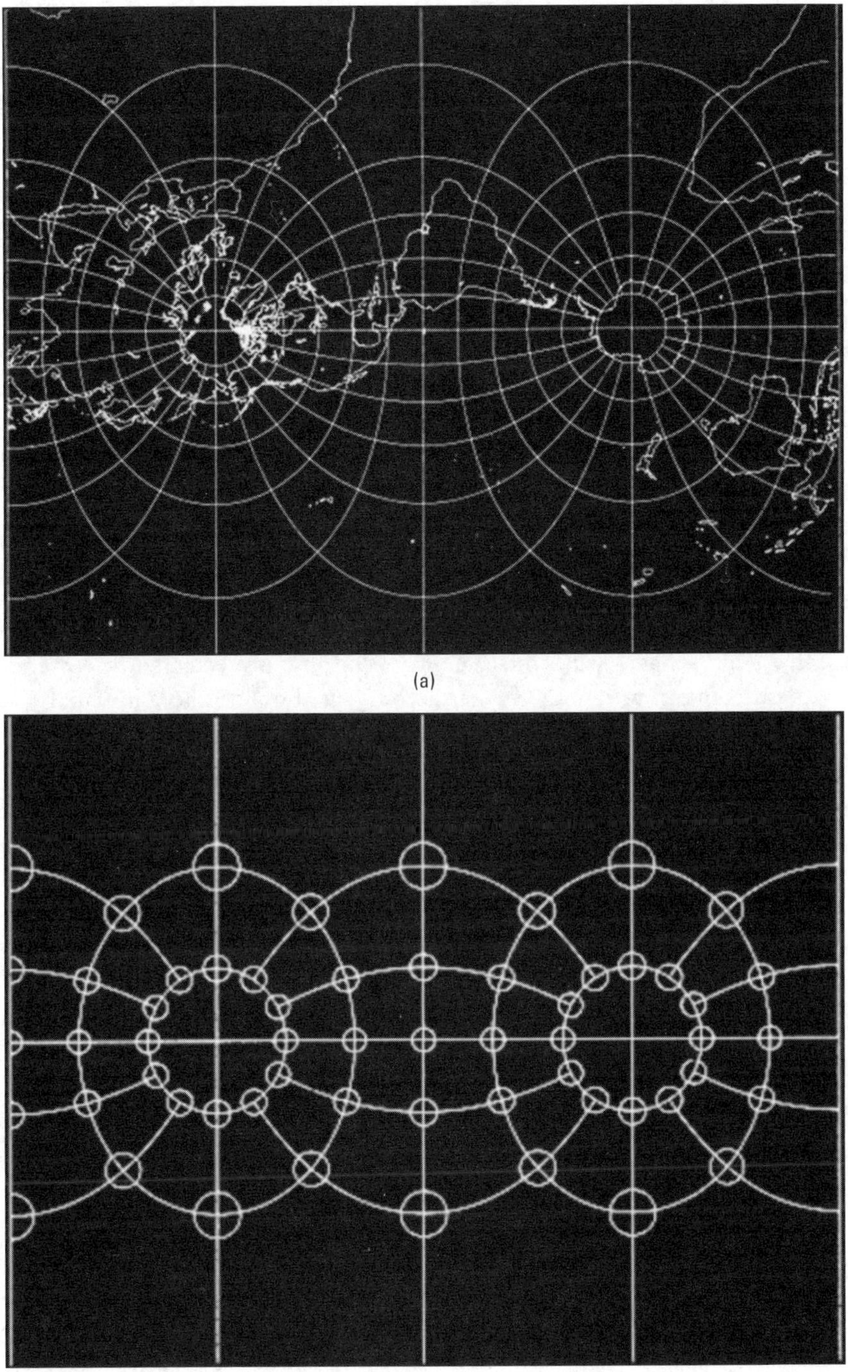

(a)

(b)

Figure 9.20 (a) Transverse Mercator projection and (b) its Tissot indicatrix.

Methods for Distortion Management

We have discussed map projection distortion in great detail within this chapter but have mildly addressed methods for distortion management. The following methods are tried and true ways to manipulate and manage distortion within any geospatial projection.

Redistribution of the Scale Factor

A proven method of manipulating the effects of and controlling the location of distortion is to modify or redistribute the scale factor. As discussed in detail previously in the chapter, all dimensions of the Earth are changed proportionately when the Earth is reduced from actual dimensions to a globe model. As the surface of this generating globe is further transformed into a projected surface, some dimensions remain unchanged. For example, in a cylindrical projection, as the cylinder is unrolled and developed, the equator maintains its original length and true scale. This *standard line* or line of exact scale is represented by a true scale factor value of 1.0.

The scale factor on any projected surface can vary at any location, since no projected surface has a uniform scale. Scale within a projected surface may also vary at different *directions* at the same location. For instance, in a particular conformal projection the scale is 1.0 (true) along each standard parallel, while other points between the standard parallels are at a scale *less than* 1.0. Accordingly, points outside the standard parallels are at a scale *greater than* 1.0. The resulting conformal projection creates a continuous error surface, in which any line (other than a standard parallel) within the plane has a unique scale factor and any line of substantial distance crosses many different scale factors.

Proportional to the original object, scale factor has innate characteristics that are linear in nature. For example if a line is doubled in length by the projection in the output surface, the scale factor of the projection is 2.0.

A basic formula to use when calculating the scale factor for distortion manipulation is

$$Scale\ factor = \frac{Distance\ on\ the\ projected\ surface}{Distance\ on\ the\ spheroid}$$

Distortion can be manipulated by redistributing the scale factors of the standard lines. The shorter the distance between the standard lines, the lower the distortion, and conversely, the longer the distance, the greater the distortion. A good implementation of this method is to position your particular area of interest in between two standard parallel lines. The projected area benefits from the true scale factor of the standard lines, while distortion is significantly reduced within the projected area.

Changing Projection Aspect

A change in the projection aspect is the most obvious and fundamental method to minimize unwanted distortion. Simply put, it is a change in the location of the projection's center point. When a projection aspect is changed, the distortion pattern in the projected map space is modified whereby areas of least distortion are relocated. An adept user can determine which projection aspect offers the most advantageous distortion relocation for the specific application.

Other Successful Methods

Aside from the two most prominent projection distortion management techniques, there are alternative methods that have proved adequate for particular applications. Here are some fine methods to employ:

- *Visual-logical analysis.* A visual examination of graticule patterns is useful for identifying local distortions (anisotropy).
- *Familiar shapes.* Recognizable features, such as coastline shapes or inland characteristics, provide important clues to distortion patterns and imbue an overall impression of major distortion points.
- *Color differentiation.* Mapping three color scales at locations within the developable surface provides insight into the projection distortion patterns. Color scales are imposed upon the *x, y* range and angular convergence parameters. Prominent differentiation between color scales offers a clear layout of definable distortion points. This color differentiation method lends itself to the structure of raster data.
- *Isarithms (isolines).* The isarithm method provides a good vantage of the overall distortion patterns in a projection. Isolines are basic contour lines that identify the differing plateaus of distortion within the projection. The result can be likened to a topographical map with designations not of elevation but of distortion magnitude. This is particularly helpful in the analysis of equal-angular or equal-area distortion. However, like the indicatrix, isarithms are static objects that are unable to accommodate the diversity of continuous application.

As we have discussed throughout this chapter, a well-planned and adeptly created map projection is critical for the basis upon which to build a GIS. In fact, the success or failure of the application resides initially with the projection. A good understanding of the final application and a basic command of projection elements will ensure a solid GIS foundation.

10

The Coordinate System

Up to this point, we have only touched the surface of coordinate systems, both in theory and function. In the past chapters, especially the chapter on map projections, coordinate systems have taken a nominal role in support of the main topic. We have already heard the likes of geographic and plane coordinate systems but have discussed very little about what these systems do. Concluding this book's theory-intensive Part II to understand the core GIS environment, we will explore coordinate systems and the role they play in a GIS.

What Is a Coordinate System?

A *coordinate* is a number set that denotes a specific location within a reference system. Typical coordinates are the *x*-*y* set ([*x*, *y*]), which is used in a two-dimensional system, and the *x*-*y*-*z* set ([*x*, *y*, *z*]), which is used in a three-dimensional system. With that stated, a coordinate system is the reference system upon which coordinates are defined. A coordinate system is often structured in a two-dimensional or three-dimensional plane that consists of a set of reference points and rules that define the spatial position of points.

In reality, coordinate systems are extensions of mathematics, particularly algebra and trigonometry, and come in various formats. Planar systems have *x*- and *y*-axes, while three-dimensional systems have an additional *z*-axis for height. Terrestrial coordinate systems, such as the ones on geographic maps, utilize a system of latitudinal and longitudinal coordinates. Polar reference systems utilize positional coordinates consisting of the distance from the origin *r* and its angle of inclination Θ. Ultimately, each coordinate system format is used on a specific type of plane.

Coordinate reference systems can also be created for specific regions or particular applications. For instance, the New York City is considered a structured, planned city, whereby streets and avenues intersect, forming a defined grid. Essentially, this is a citywide reference system upon which positions can be easily defined and understood. You can say your position is at 44th Street and 6th Avenue and immediately your position is understood within the boundaries of the city.

This same premise goes for road maps, which commonly have a zoned grid defined by horizontal numbering and vertical lettering. These zones do not define exact positions, but a grid rectangle (area) in which the target resides and the locator can search. For example, you can find a particular town by locating and searching its particular zone (i.e., A6). If there are multiple page maps, the zone must be detailed with the page number (i.e., page 6, A6) since most multipage maps use the same grid identifiers on each page. This procedure of "zeroing in" on a target's position dramatically reduces search times and is comparable to stating the city street you are on (zone) and not the actual address (position).

Mathematics and the physical Earth require the use of various standard coordinate systems. In relation to GIS and geodesy, there are a handful of important reference systems worthy of discussion and necessary for the novice GIS user to understand. These coordinate systems include the Earth-based geographic coordinate system, the vector-based Cartesian coordinate systems (both two-dimensional and three-dimensional versions), the zone-based Universal Transverse Mercator and Universal Polar Stereographic systems, and the U.S.-based State Plane Coordinate System. Let us take a brief look at each.

The Geographic Coordinate System

A geographic coordinate system is a three-dimensional positional reference that utilizes latitude, longitude, and ellipsoidal height. We have briefly explored the geographic coordinate system in our map projection discussion. This positional reference system is among the most used today for global locations and is typically associated with geodetic datums.

To recap, the Earth's three-dimensional ellipsoid is mapped through a series of horizontal and vertical reference lines, which form a set of standard reference lines. As we know, the horizontal lines are called *longitude* lines. Longitude lines are parallel with one another and are therefore often called *parallels.* The vertical lines are called *latitude* lines. Since latitude lines meet at the poles, they are often called *meridians.* Figure 10.1 illustrates Earth's latitude and longitude lines.

Earth's parallels intersect every meridian at right angles. The central parallel, located exactly between the poles, is called the equator. This central,

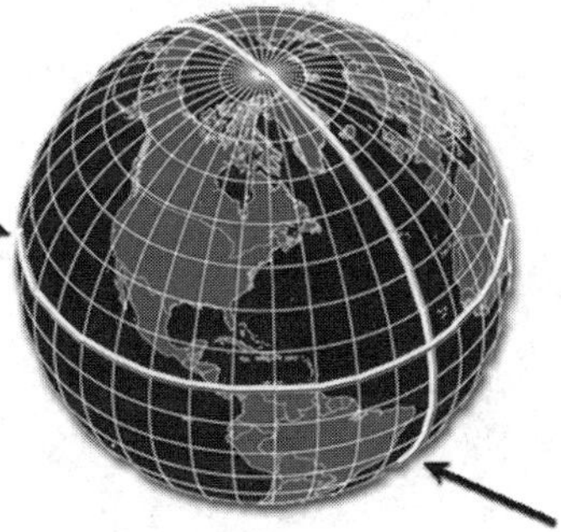

Figure 10.1 Earth's latitude and longitude.

standard parallel holds a longitudinal value of 0°. The remaining parallels are assigned angular values relative to their direction and distance from the equator. The lines of longitude progressing toward the north pole are assigned parallels ranging from 0° to 90° north. Likewise, the lines of longitude progressing from the equator toward the south pole are assigned parallels ranging from 0° to 90° south.

Earth's meridians run from pole to pole, and unlike Earth's parallels, meridians convene at each pole. The center meridian is called the prime meridian, which is the line of latitude that crosses through Greenwich in the United Kingdom. The prime meridian is the primary latitude reference line and is assigned an angular value of 0°. The remaining meridians are assigned angular values relative to their direction and distance from the prime meridian. Meridians progressing east of this center meridian are assigned angular values ranging from 1° to 180° east. Similarly, the lines of latitude progressing west of the center meridian are assigned angular values ranging from 1° to 180° west.

The geographic coordinate system uses geodetic latitude and longitude to define position on an ellipsoidal surface and, as such, to form the geographic coordinates. Geodetic latitude and longitude are identical to the Earth-based system. In addition, the geographic coordinate system provides a quantity for geodetic height. Geodetic height at a point is the distance from the reference ellipsoid to the point in a direction normal to the ellipsoid. Figure 10.2 illustrates the geodetic latitude, longitude, and height.

The geographic coordinate system provides a solid reference through which any location on Earth can be uniquely distinguished and identified. As we have seen with map projections, GIS generally transforms three-dimensional geographic coordinates to other coordinate systems in order to clearly portray point positions on a two-dimensional information product. This leads us to our second coordinate reference system, namely, the Cartesian coordinate system.

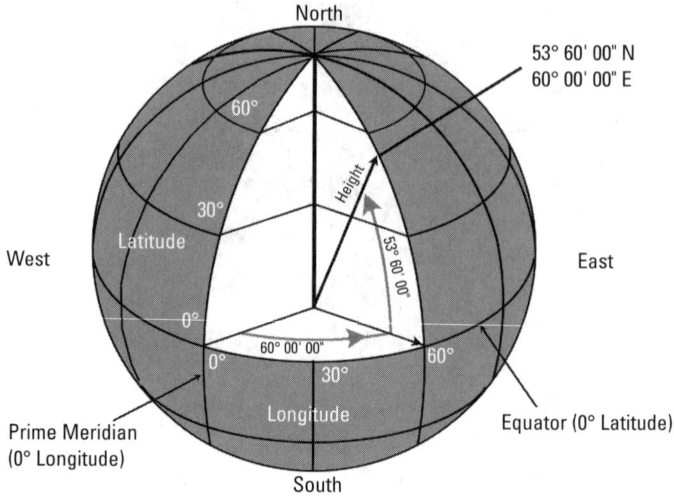

Figure 10.2 Geodetic latitude, longitude, and height.

The Cartesian Coordinate System

Perhaps the most used coordinate reference system in mathematics, science, and GIS is the Cartesian (or rectangular) coordinate system. The Cartesian system is a reference structure in which point positions are measured along intersecting planes in two and three dimensions. The common coordinate or intersection point for all planes is called the origin. All measurement distance increments are consistent on every intersecting plane and throughout the system. These planes are detailed along the x, y, and/or z axes.

The Cartesian coordinate system was named for famous French mathematician, scientist, and philosopher René Descartes (1596–1650). Descartes, known also as Cartesius, is often dubbed the "Father of Modern Mathematics" for his mathematical genius and his creation of analytic geometry. In his *Discourse on the Method to Rightly Conduct the Reason and Search for the Truth in Sciences* (usually shortened to *Discourse on Method)*, Descartes detailed his method of translating geometric problems into algebra using a coordinate system. This mathematical innovation paved the way for later mathematical advances, particularly calculus.

The Cartesian two-dimensional coordinate system involves two axes: the horizontal x axis and the vertical y axis. The origin is detailed as coordinate (0, 0). This planar, two-dimensional model is divided into four quadrants by the perpendicular, intersecting axes. These are labeled quadrants I through IV, starting in the upper right-hand quadrant and progressing counterclockwise. These quadrants take on the following structures:

- Quadrant I: *x* axis is positive, *y* axis is positive (+*x*, +*y*);
- Quadrant II: *x* axis is negative, *y* axis is positive (−x, +*y*);
- Quadrant III: *x* axis is negative, *y* axis is negative (−*x*, −*y*);
- Quadrant IV: *x* axis is positive, *y* axis is positive (+*x*, −y).

Figure 10.3 offers a good line drawing of the standard two-dimensional Cartesian coordinate system.

As you might have guessed, the Cartesian coordinate system is perfect for plotting and measuring vector-based feature geometry. A GIS references the positions of feature-defining points, lines, and polygons using the Cartesian system of coordinates. In addition, you may remember that a map projection brings three-dimensional geographic coordinates (latitude and longitude) into two-dimensional Cartesian (or planar) coordinates.

Often points are located in a three-dimensional space and a height measurement is necessary. Height is measured along a third axis—the *z* axis.

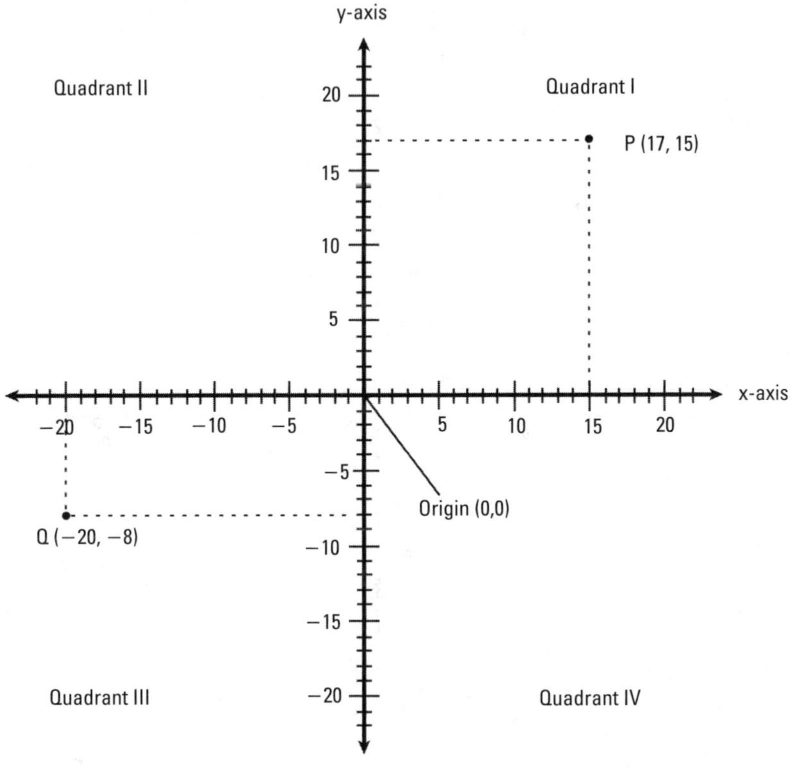

Figure 10.3 Two-dimensional Cartesian coordinate system.

Three-dimensional objects, like the Earth's ellipsoid, utilize a height measurement for positional accuracy. Consequently, geodetic datums use this system. Figure 10.4 details the three-dimensional Cartesian coordinate system.

Frequently, the Earth is set to a three-dimensional Cartesian system called the *Earth-centered, Earth-fixed* (ECEF) Cartesian coordinate system. ECEF is used to define three-dimensional positions with respect to the earth's mass center of gravity and the center of the reference ellipsoid. The Earth's center serves as the origin (0, 0, 0). You may want to refer back to Figure 10.2, which details an ECEF system as it relates to the standard lines: the equator and prime meridian.

This form of Cartesian coordinate system is especially useful for geodetic structures. It is called Earth fixed because each axis is fixed in respect to the orientation of the Earth. For example, if the Earth rotates, the axes rotate sympathetically and proportionately.

Before leaving the Cartesian system discussion, we need to explore an important terminological matter. Habitually, cartography and surveying practices use Cartesian coordinates to define positions on maps. As a matter of function and ease, the *x*-axis is considered an *easting* and the *y*-axis is considered a

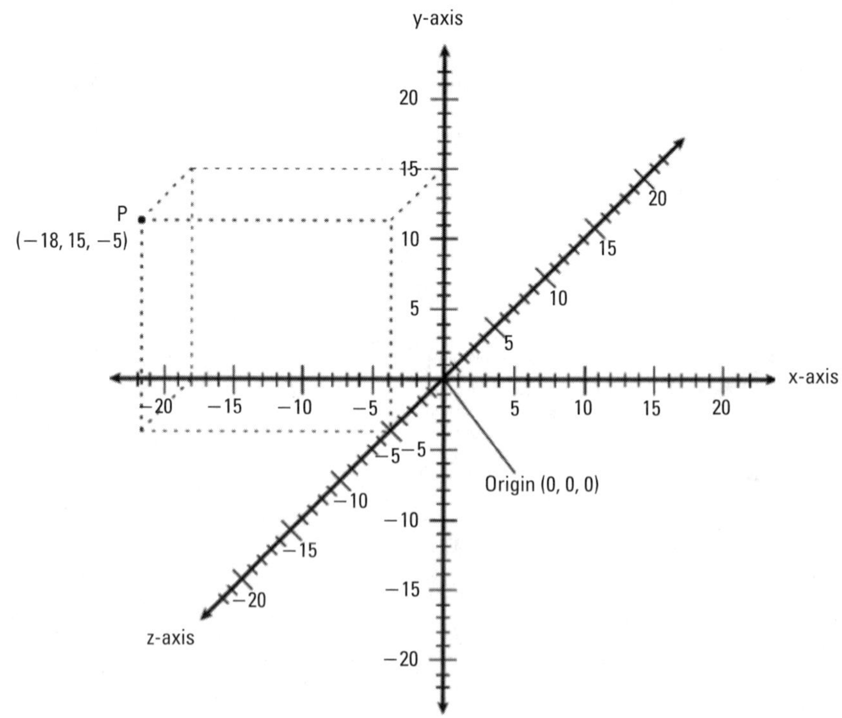

Figure 10.4 Three-dimensional Cartesian coordinate system.

northing. The northing-easting terminology is used to help efficiently relay position, while avoiding signage and coordinate errors.

An easting is the distance east from a north-south reference line, or meridian. Often the prime meridian is used as this reference line. Similarly, a northing is the distance north from an east-west reference line, or parallel. The equator is often used as this reference line.

Furthermore, to simplify positions in the northing-easting reference method, negative numbers are eliminated. To do this, *false easting* and *false northing* constants are used. A false easting is an adjustment constant added to *x*-axis coordinates to eliminate negative numbers. Likewise, a false northing is an adjustment constant added to *y*-axis coordinates to eliminate negative numbers. On maps using the northing-easting terminology, false easting and false northing values are supplied.

Universal Transverse Mercator

The Universal Transverse Mercator (UTM) is a global map projection that transforms the three-dimensional world into a two-dimensional system. The UTM employs an international plane coordinate system that extends around the world from 84° north above the equator to 80° south below. It is extended an extra 4° in the North to cover the northernmost land on Earth. This bounded coverage is because the UTM projection distorts near the north and south poles. The Earth's polar regions are handled by the Universal Polar Stereographic projection, which is discussed later in this chapter.

The UTM coordinate system is set upon a zoned grid, which divides the Earth into 60 equal zones that are all 6° wide in longitude (east-west). The UTM zones are numbered 1 through 60, starting at the international date line (zone 1 at 180° west longitude), progressing east past the prime meridian (zone 30), and back to the international date line (zone 60 at 180° east longitude). Figure 10.5 illustrates the complete UTM zones.

As depicted in the figure, each zone extends in both the northern and southern hemispheres and employs the northing-easting methodology for position identification. Since the equator separates the hemispheres, it serves as the zone's standard parallel. Each UTM zone refers to its own central meridian located at each zone's middle (3°).

Given that the northing-easting method does not allow for negatives, the equator separates each zone into two positive sections: a north section and a south section. Each section then employs its own rectangular grid system using the equator and the zone's central meridian as the two standard references for each system. In total, the UTM comprises 120 separate coordinate systems.

Coordinate positions referencing the UTM system must be indicated with the zone number as well as its northing and easting values in meters. The easting

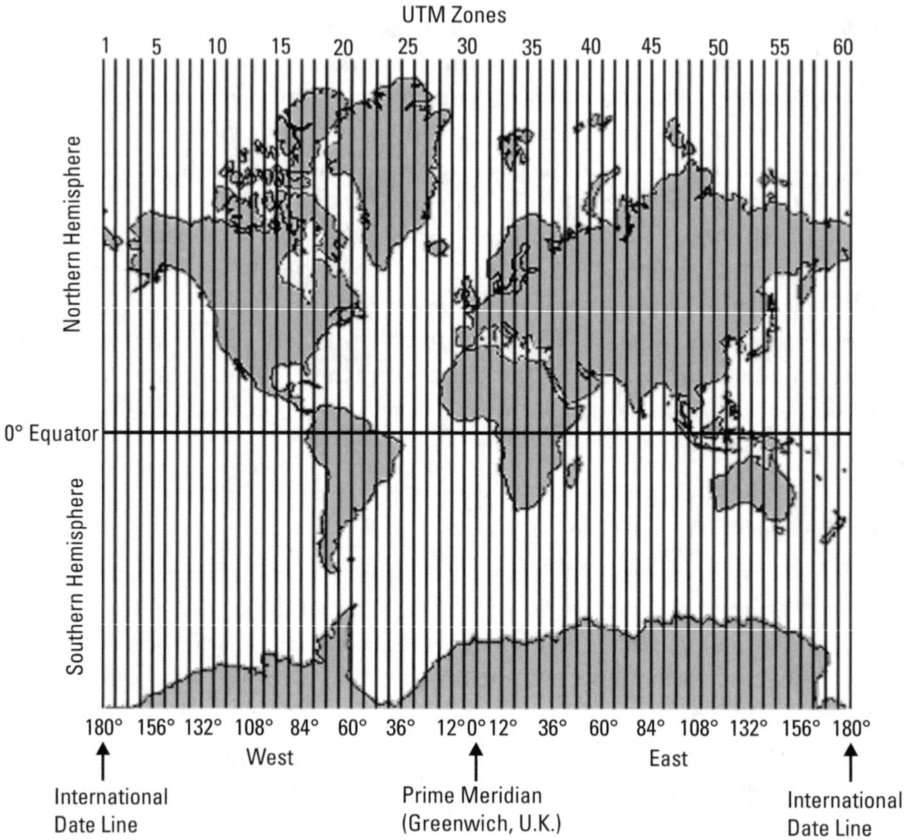

Figure 10.5 The Universal Transverse Mercator zones.

value is always a six-digit value in meters, while the northing value is nearly always a seven-digit number in meters. For example, a typical UTM coordinate for zone 13 would look like *13 645,000mE 2,870,000mN*. Furthermore, to ensure positive coordinates, each UTM zone utilizes a false northing value of 0m in the northern hemisphere and 10,000,000m in the southern hemisphere, and a false easting value of 500,000m.

The UTM system is used worldwide for a multitude of applications, although it may not always be an identical system. For example, in Europe, the Kruger-Gauss projection is used, and the system is consequently called the Kruger-Gauss Transverse Mercator. In the United States, the military uses a modified UTM system called the Military Grid Reference System (MGRS), which further divides the UTM zones into 8° latitudinal divisions (the last division with 12°) designated with a letter. The MGRS starts at C at the 80° south latitude and progresses to X ending at the 84° north latitude. The X latitudinal division is the only division 12° high. Additionally, to avoid user error, the letters I and O are not used because of their similarity with numbers.

In summary, the UTM is a widely used planar coordinate system quite capable of representing the Earth's geographic coordinates on a flat, northerly-easterly plane. The UTM is used worldwide, and in many countries it serves as a basis for their national grid, particularly the United States and the United Kingdom. Incidentally, the approximate scale distortion of the UTM system is only 1 part in 2,000, proving that the UTM is a highly accurate system. Still, as mentioned earlier, the UTM projection cannot be used for the Earth's polar regions due to amplified distortion. Let us now examine UTM's corresponding system that focuses on theses polar regions—the Universal Polar Stereographic.

The Universal Polar Stereographic

Serving as the polar version of the UTM, the Universal Polar Stereographic (UPS) system uses a universal polar stereographic projection to transform the three-dimensional polar regions into a two-dimensional system. The UPS takes over where the UTM projection distorts, which is defined as above 84° north latitude and below 80° south latitude, but also includes an additional 30' of latitude, extending into the UTM grid for a small degree of overlap between both systems. Figure 10.6 portrays the approximate combined geographic coverage of Earth through the UTM and the UPS systems.

As depicted, the UPS system has two zones, namely, the north and south polar zones. Like the UTM, each zone has its own grid structure and coordinate system. Northings and eastings are used to characterize coordinate positions for each zone and are subsequently calculated using a polar aspect stereographic projection. Figure 10.7 presents a polar aspect stereographic projection.

Similar to Figure 10.7, the UPS system employs a conformal map projection that is free of angular distortion. The projection retains a constant scale along the parallels and is true scale at 87° 7′ north and south. As with the UTM, the U.S. military utilizes a gridded version of the UPS system in which each grid block is identified with a distinguishing two-letter combination (e.g., ZA, where the easting coordinate is defined by Z and the northing coordinate is defined by A).

Overall, the UTM and UPS systems are complementary to one another and combined provide a whole flat view of Earth. The zoned, multigrid approach offers a wonderful positional mapping capability to cartography and surveying, while the minimized distortion provides a near true-to-life representation. In short, the UTM-UPS combined system is a good way to characterize the Earth as a whole and to lay a foundation for both coordinate representation and measurement.

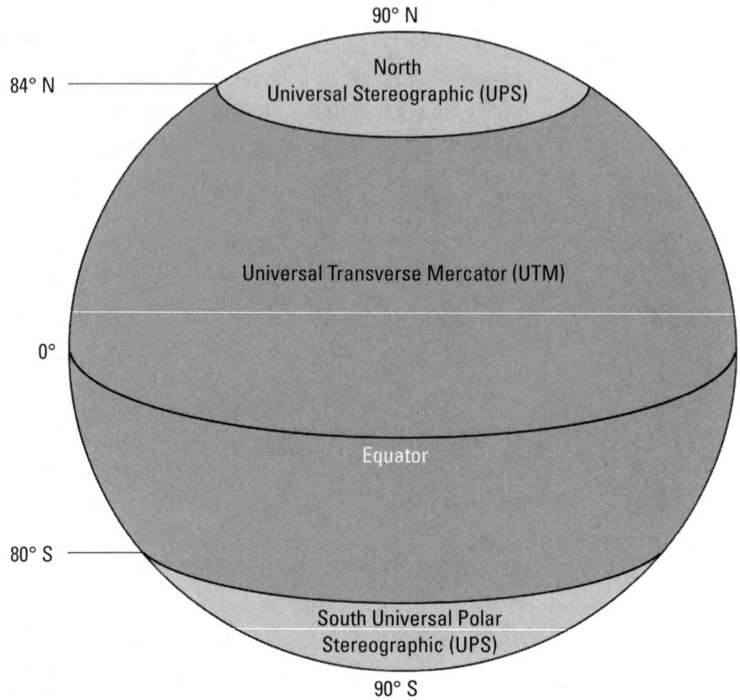

Figure 10.6 The UTM-UPS global coverage.

The State Plane Coordinate System

The State Plane Coordinate System (SPCS) is a zoned grid system that divides all 50 United States into more than 120 numbered zones. The SPCS was first introduced in the 1930s by the U.S. Coast and Geodetic Survey to serve as a common reference for cartographers, surveyors, and engineers. The SPCS is used only in the United States and allows coordinates from an entire state or portions of a state to be transformed into planar (rectangular) coordinates on a single grid. The SPCS was developed to have even greater precision than the UTM, with its maximum scale distortion at just 1 part in 10,000.

Unlike the UTM system that covered both sections of the northern and southern hemispheres within one zone, SPCS zones are bounded by state borders. Smaller states are divided into one or two zones, while larger states are divided into as many as six zones. Each zone has its own origin, central meridian, and standard parallel which often follow state county boundaries. Additionally, for easy identification, each zone has its own assigned code.

Figure 10.8 illustrates the SPCS used in Arizona. It is important to recognize that the state is divided into three zones, which are attributed for their

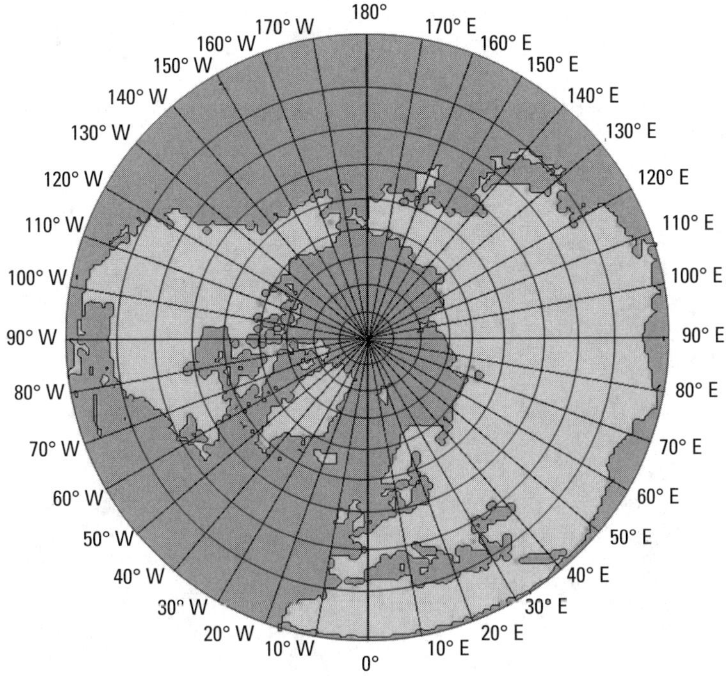

Figure 10.7 A north polar aspect stereographic projection of the Earth.

spatial location within the state: Arizona West, Arizona Central, and Arizona East. Each zone is assigned a unique four-digit code encompassing a common state prefix 02 and an incremental two-digit zone suffix. As illustrated, Arizona's zones follow the state's county lines.

The original SPCS was developed using the standard national datum NAD27 (North American Datum of 1927). Appropriately called the U.S. State Plane Coordinate System of 1927 (SPCS27), this early version references the Clarke 1866 ellipsoid and creates a set of local grids for each state. Like NAD27, the SPCS27 system is based on the foot measurement—a U.S. standard that has since changed.

After NAD27 was superseded by the more accurate NAD83, the SPCS was revised. The new system, similarly called the U.S. State Plane Coordinate System of 1983 (SPCS83), is based upon the latest national standard horizontal datum (NAD83). SPCS83 moved away from the antiquated foot measurement to the U.S. standard, the meter. Due to its NAD83 foundation, SPCS83 offers greater geodetic control and cross-links to the GPS-based GRS80 datum.

The SPCS is based upon two map projections: the transverse Mercator and the Lambert conformal conic. For states that are elongated north to south,

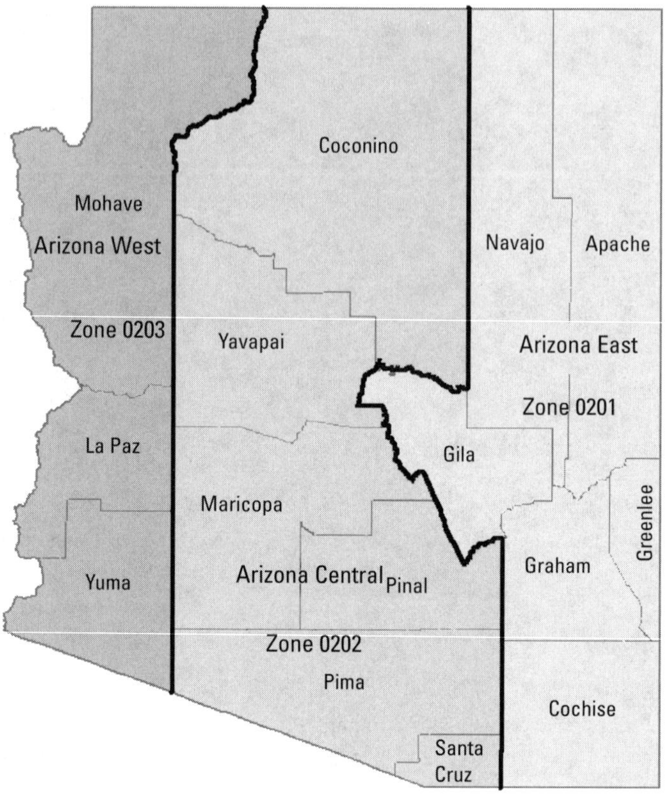

Figure 10.8 The State Plane Coordinate System in Arizona.

such as California and Florida, the SPCS utilizes the Lambert conformal conic for minimal projection distortion. For states that are elongated east to west, such as Pennsylvania and New York, the SPCS utilizes the transverse Mercator.

The SPCS also utilizes the northings-eastings method of coordinate identification. Like other planar coordinate systems, false northings and false eastings are used to keep coordinates positive for every zone. Since every zone is slightly different, each has its own values for false northing and false easting. Overall, the SPCS system is an exceptionally precise and easy system to employ throughout the entire United States.

Coordinate System Implementation in GIS

Coordinate systems play an important role within GISs. They are used throughout every form of geodetics from geographic information referencing and vector feature geometry to geodetic map projections and datums. The coordinate system forms the definitive core of GIS and serves as the primary tool for modern geodetics.

It is important to understand that the implementation of a coordinate system in a GIS enables geographic data to be defined by vector data. The positional accuracy of vector data on a coordinate system allows for real-world features to be precisely represented in the GIS, as well as on the ever-important geographic information products. Vector data also enables feature topology and its resulting, advantageous efficiency. In short, GIS accuracy and vector data availability are both direct products of utilizing a coordinate system.

Another beneficial effect of coordinate system implementation is that geodata from various (different) sources can be used with one another through layering techniques and coordinate system transformations. By knowing the source data's reference system, feature coordinates can be converted to a single, uniform GIS coordinate system. With all the data sources converted, the information can be manipulated on the same grid, and associations between sources can be made. It can be noted that without the implementation of a coordinate system, a GIS would remain just a database.

Part III:
GIS Applications and Environments

Geographic Data for Analysis and Presentation

11

Thematic Mapping

In his published anecdotes, Samuel Johnson, noted eighteenth-century writer and philosopher, wrote, "The use of traveling is to regulate imagination by reality, and instead of thinking how things may be, to see them as they are." We are those travelers in that, throughout this book's final part, we will be traveling from GIS's theory into the territories of true GIS products and real-life applications. As part of this demystification of GIS, we will depart from discussing how information products should be (imagination) to how they really are (reality).

We already know that in order to produce useable geographic information products, the user must manipulate geodetics through the selection of a standard ellipsoid and datum for their particular focal area. We also know that, once selected, a beneficial map projection must be chosen and applied upon a defined coordinate system. After these elements are set, valid geographic information products can be created and analysis can be conducted with a degree of quantified certainty.

One of the most common and valuable products a user can produce with GIS is a map. Some map products depict pure cartographic information. These products include road maps, subway maps, and the like. Other map products offer a mechanism for comprehensive geographic data analysis and, as such, are typically more valuable. These specialized maps utilize user selection and the benefits of a GIS to offer a unique, yet comprehensive, description of the focal geographic terrain. This specialized map product is called a *thematic* map.

The Need for Thematic Mapping

We have all seen and used a thematic map of some sort. Think of any major city subway or bus map with the stations or stops depicted, or of a TV weather map with temperatures and high- and low-pressure areas represented. Well, the stations, stops, temperatures, and highs and lows are all attributes of geospatial data. These dataset attributes are called themes, and subsequently the subway, bus, and weather maps are all variations of thematic maps.

A thematic map is defined as a specialized map that depicts the spatial distribution of one or more explicit themes. These map themes portray dataset attributes and can detail anything quantifiable, such as population density, income distribution, crime, natural resources, water distribution, pipelines, and disease occurrence. Since GISs are used to manipulate, analyze, and convey geospatial information, thematic maps offer the GIS user a better way to understand the geographic area in focus, as well as an excellent means to inform others.

Now, just imagine how difficult it would be for the user if the bus stops were identified in a data table rather than graphically on the map or if the temperatures for 20 neighboring locations were spoken by the broadcaster and not represented on the map. It is quite possible that some commuters would miss their stops, while the comprehension of temperatures for 20 locations would become an exhausting test of mnemonics and listening. Without thematic, visual indicators, the delivery of this data would be less than effective.

Thematic maps offer a means to define data attributes in a way that is not fully available from a data table. In fact, thematic mapping offers the GIS user a better instrument to convey geographic information and geospatial data that is relative to their information objectives. This, in part, provides a greater means to analyze data sets and find trends and spatial patterns that may remain hidden in data tables. Figure 11.1 offers some insight into how a data table from the U.S. Census Bureau can be interpreted differently when portrayed as a thematic map.

As demonstrated in Figure 11.1, the data table containing household and county information for the state of Florida "hides" the housing trends that the map freely depicts. One definitive trend, in particular, is the state's household per county ratio, whereby the highest ratio is in southeastern Florida and the lowest ratio in northern Florida. The thematic map helps visualize these trends and spatial patterns, while the data table, although informative, seemingly obscures Florida's popular household areas within its data.

Thematic maps offer a bird's-eye view of a particular attribute within the context of a geographic area. As such, they can provide either qualitative information or quantitative information. Examples of qualitative information are vegetation types, soil conditions, and land use categories. Quantitative information includes such attributes as number of Hispanic homeowners, carbon

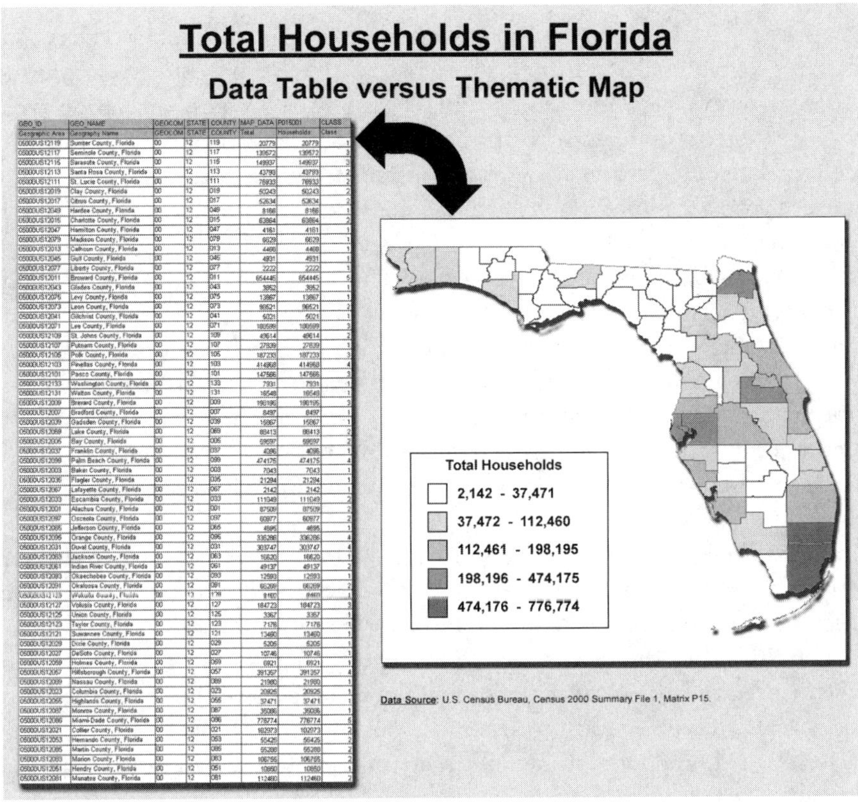

Figure 11.1 A data table versus a thematic map.

monoxide emission measurements, and soil depths. The thematic map can depict both qualitative and quantitative data only if two or more themes are used in tandem.

In the end, the absolute need for well-designed thematic maps becomes apparent whenever medium to complex attribute analysis is undertaken. Since users have inherent limitations to their mental acuity, visual models are necessary for enhancing data comprehension and vitalizing the data's analytic potential.

Manipulating Thematic Layers

A GIS offers users an easy method to manipulate specific data attributes through the application of thematic layers. A layer is a set of theme-based data that is described and stored in a map's data library. Layers do not contain geographic data, per se, but are georeferenced to geographic data. Additionally, layers complement geographic data through a basemap or satellite image, or such information as roadways, pipelines, soil sample locations, soil types, vegetation types, and water distribution. Layers can be turned on or off in a GIS and can be ordered relative to viewing priority. For instance, a higher priority theme would be placed atop a less important theme, such as a railroad track layer atop a background map of the area.

Figure 11.2 offers a good, visual explanation of thematic layers. The composed thematic map is a combination of five thematic layers: cities, railroads, interstates, major roads, and states. These layers are logically ordered for best overall viewing, with the attribute layers overlying the background layer (states). If the states layer had the highest priority for viewing, then it would block the other themes from showing through. Therefore, part of mapmaking with a GIS is both realizing the importance of logical layer order and keeping data conveyance at a premium.

For GIS users, thematic layers prove to be incredibly useful when developing information products. What makes thematic layers so advantageous is the fact that they provide data attributes in the form of features and surfaces that are uniquely linked to the geography. These features and surfaces are layered and georeferenced to help advance the overall understanding of the terrain and offer insight into what objects affect the landscape, and to what degree. Incidentally, this forms the foundation of thematic mapping's enhanced geospatial analysis.

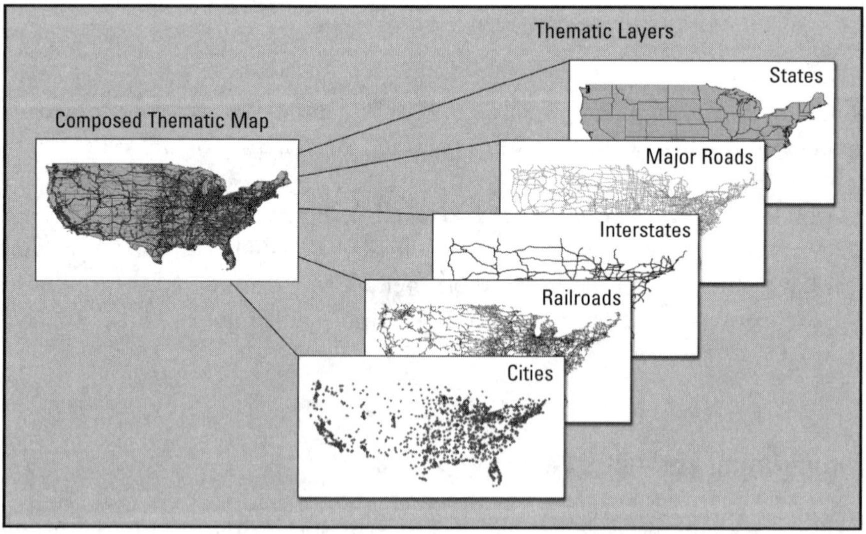

Figure 11.2 How thematic layers constitute a thematic map.

Thematic layer *features* are either natural or manufactured objects that are present or exposed to the terrain, either above ground or below ground. Thematic features include natural objects, such as wetlands, mudflats, and hydrology, or manufactured objects, such as pipelines, roadways, and electrical distribution lines. Each map feature has a defined geospatial position, shape (i.e., point, polygon, or line), and identifier (or symbol) that represents one or more of its characteristics.

Thematic layer *surfaces*, on the other hand, are slightly different and offer a different type of information. These layer surfaces offer a perspective to the terrain in terms of specific characteristics. Examples of thematic surfaces include elevations, soil density, temperature, pressure, precipitation, and flooding. Unlike features, surfaces do not always have defined shapes.

Proper thematic mapping involves a certain level of user discipline relative to thematic layers, as well as a consistent layering scheme. For effective GIS use and map control, layers should be consistently named, with easily identifiable labels. These layer names should suggest the layer's function and indicate some type of hierarchy. Suppose you are entitling a thematic layer of regional streams. A title such as "rty65sc" would make a poor thematic layer name, since it is not easily ascribed to streams or anything else. A better choice would be "Streams001" since it is obvious and has a method to define a stream hierarchy.

Now imagine having 20 different thematic layers for a particular region, including seven stream-related layers, six soil-related layers, and seven layers defining various utilities (gas, electric, and so forth). Which label nomenclature would you rather use? What if you had to produce 22 multithemed maps of that region for a client? Without a disciplined approach to layer taxonomy, this thematic mapping undertaking can rapidly become an onerous task.

As with everything in GIS, there is no single correct way to manipulate or display thematic information. GIS is all about personal choice, creativity, and user flexibility. As such, GIS offers various thematic mapping types from which to choose. We will take a closer look at each thematic map style to understand how and when each is appropriate.

Types of Thematic Maps

GIS offers a fair degree of flexibility and choice when it pertains to what geospatial information is used, how it is used, and the manner in which it is presented. Similar control is offered when depicting attribute data in a thematic map. Consequently, there are numerous ways thematic information can be portrayed over a geographic region. Depending upon the particular data portrayal or analytic objective, the GIS user has a certain amount of freedom over the mapping design.

Given the widespread implementation of thematic mapping, four prominent map types have surfaced. Each offers a particular analytic benefit and, as mentioned previously, can depict qualitative or quantitative data. These four primary thematic map types are choropleth maps, graduated symbol maps, dot density maps, and isopleth maps. Let's examine each one.

Choropleth Maps

Perhaps the most prevalent type of thematic map, the *choropleth map* offers a definite method of portraying ratios, densities, or proportions aggregated within a specific region. These regions are defined by bounded areas, such as census tracts, counties, states, or countries. Choropleth maps, also known as *graduated color maps,* depict thematic information in these bounded areas through color shading and color/value attribution.

Figure 11.3 provides a good example of a choropleth map and illustrates how state median age data can be portrayed in grayscale. The map quickly offers spatial patterns associated with age and individual states that would have been nearly impossible to discern from the data alone. Notice how the eastern states have a higher median age, while the central states hover closer to the middle of the age spectrum. It is easy to visually identify that Utah has the lowest age group (youngest median population).

A choropleth map assigns a value or range of values (called *data classes*) to a specific color or shading. In the figure, the data classes were assigned to five different shades of gray. Hence, the figure has five data classes. For best results,

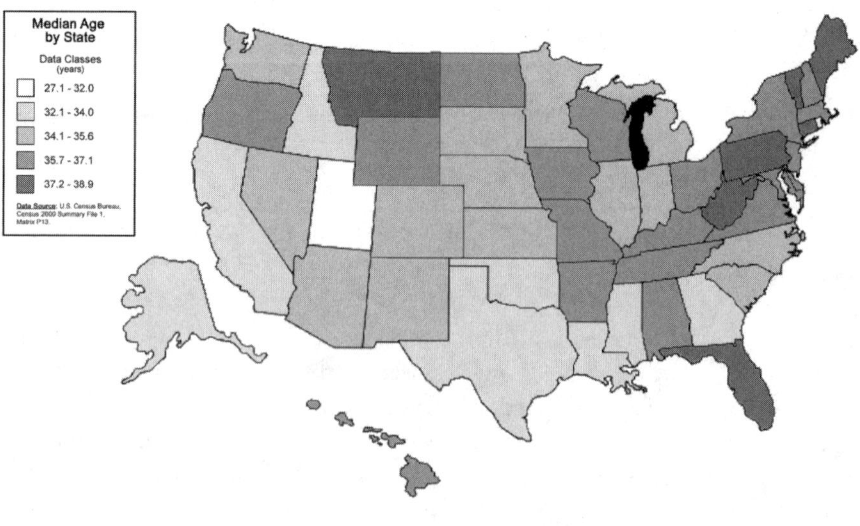

Figure 11.3 A choropleth map of the U.S. census data on median age by state.

choropleth maps should utilize no more than eight data classes since most individuals have a hard time distinguishing between many subtle variations in colors.

Choropleth maps work best with discrete data over defined, bounded areas. Problems arise when choropleth maps are detailed with too many unique values attributed by unique colors. In such cases, the map turns into a geographic rainbow and loses its analytical worth. Poor map designs lead to spatial patterns and communal trends that get lost in the sea of differentiated color.

Choropleth colors for data classes follow particular conventions, whereby the rule of thumb is: The lighter the color, the smaller the value; the darker the color, the greater the value. In the median age map (Figure 11.3), the color white indicates the youngest age range (smallest value), while the darkest gray indicates the oldest age range (greatest value). These maps are portrayed in grayscale or color, typically using a predefined color scheme called a *color ramp.* The grayscale colors in the figure form a predefined color ramp. Users also have the option to bypass the color ramp scheme to define their own color-data class attribution. Whatever color scheme you use, the overall objective is to provide choropleth maps that are simple to understand and offer the greatest potential for spatial trend analysis.

Graduated Symbol Maps

The second type of map is the *graduated symbol map,* which is similar to choropleth maps in function but uses symbols (typically a circle or dot) instead of a color fill. Graduated symbol maps are highly beneficial when indicating quantity or magnitude of a certain phenomenon in relation to geographic positioning. These thematic maps indicate positional data and attribute values through a common symbol of varying size. As the choropleth had a rule for its colors, graduated symbol maps have a similar rule of thumb: The smaller the symbol, the smaller the value; the larger the symbol, the larger the value. Figure 11.4 offers a sample graduated symbol map that details total population by U.S. state.

In Figure 11.4, state populations are shown in relative approximation using various sizes of dots. Each size is assigned a specific value (e.g., 1 million) and is placed in the appropriate states. You can easily recognize a pattern for the country's population through this particular map type, with the east coast having more states with a greater populations than the west coast.

Aside from the "standard" graduated symbol map, a few variations exist for enhanced data portrayal. One variation is called the proportional circle map, wherein the map symbols (graduated dots) are proportionally sized according to their represented value. For example, a proportional circle map of the national population (i.e., Figure 11.4) would perhaps have a miniature dot to represent 1

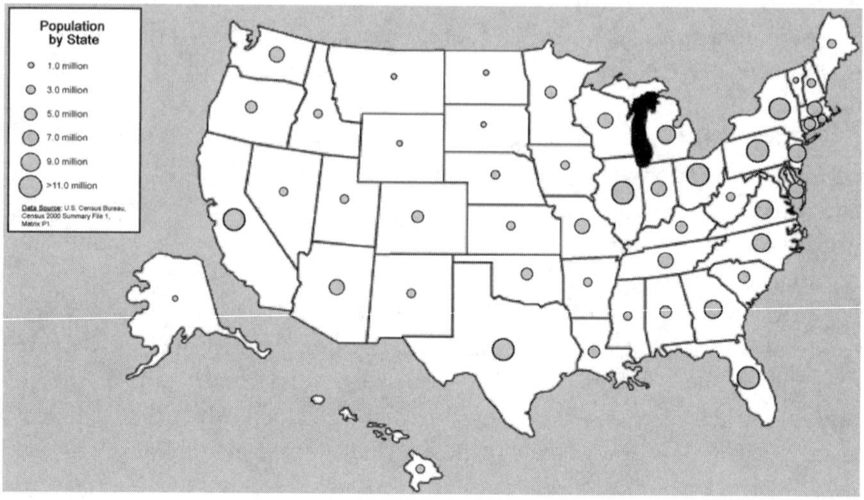

Figure 11.4 A graduated symbol map of U.S. state populations.

million people, a dot that is 4.5% larger to represent 4.5 million, and another dot that is 10% larger than the miniature dot to represent 10 million people. Value and symbol size are duly proportional.

Another variation is the pie chart map, wherein the symbol used to represent data is a proportional pie slice. This type of map depicts two pieces of discrete data for each bounded region (e.g., state). Last, a third variation similar to the pie chart map is the bar/column chart map. This map portrays bar graphs as a symbol for each region or state and, like its pie chart counterpart, offers a way to depict two forms of discrete data. Ultimately, the most popular of all the variations of graduated symbol maps is the "standard" version.

Dot Density Maps

Often, there is a need to show theme concentrations over a defined region. Choropleth and graduated dot maps are ineffective in the conveyance of thematic densities when they apply to specific geographic locations (as opposed to aggregate areas). For these instances, a *dot density map* is the choice. As the name implies, dot density maps depict spatial densities of particular thematic elements or events.

These specialized thematic maps are excellent for discerning area-theme relationships and can provide insight into the concentrated patterns of a bounded area (e.g., a state). Dots on the map are always uniform in size. Unlike other thematic map types, each dot (indicator) on the map represents a defined number of thematic elements for a specific geographic position.

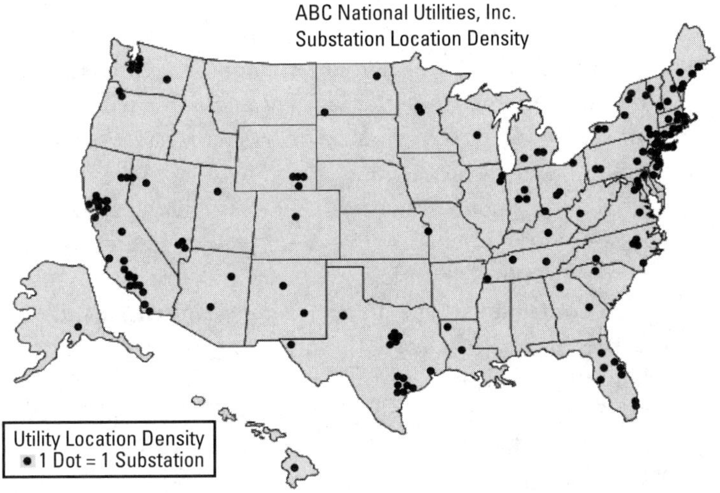

Figure 11.5 Dot density map.

The dot density map portrayed in Figure 11.5 offers an example of a typical theme density map and a perspective on how the thematic densities assist dataset analysis. Just glancing at the map provides the user with important substation concentrations throughout the United States. Ideally, it could help ABC National Utilities understand where it has an abundance of utility facilities and where it has few or none. Each dot represents one substation and provides an approximate position within each state.

Dot density maps are sometimes problematic relative to understanding exact numbers or precise coverage. In areas of abundance, such as the New York City–New Jersey–Connecticut area on Figure 11.5, the dots occupy a good portion of the map space. Since New York City is completely covered with black dots, it can be mistaken as meaning hundreds of substation locations rather than the actual, much smaller amount. In addition, dot density maps could give the wrong impression of an overabundance in a region when the amounts are adequate for the given population and area.

Aside from these few inequalities, dot density maps offer an excellent way to depict attribute densities throughout a geographic region. They are simple to understand and effectively represent concentration patterns and spatial trends across the entire mapped terrain.

Isopleth Maps

The last type of thematic map we will discuss is called the *isopleth map.* Known also as an *isoline map,* isopleth maps visually depict continuous or nearly continuous data in contiguous patterns of color or shading. Like choropleth maps, values are broken down into data classes and are represented by a color ramp (or user-defined ramps). Isopleth maps connect same data classes together with lines that form the perimeter to same data class areas over the geographic region. The resultant map of same value areas presents a sophisticated view of the spatial patterns in the region, as well as a means to define regional trends.

Due to the manner in which continuous data are represented, isopleth maps are frequently used by scientific, environmental, and meteorological professionals. Typical continuous data mapped with isopleths includes temperature, emissions, precipitation, rainfall, elevation, and contamination. Isopleth mapping can depict a variable data feature within a thematic map or as part of a topographic map (isoline). As an example, Figure 11.6 represents a real-world isoline map created by the U.S. National Atmospheric Deposition Program (NADP) to depict the country's total precipitation in 2004.

In Figure 11.6, the continuous data has no inner boundaries other than the perimeters defined by the explicit data classes. The precipitation trends for the country are well documented in that two "hot" precipitation zones are identified: the southeastern states (i.e., the Louisiana–Alabama region) and the immediate northwest (i.e., the Washington–Oregon region). Because there are 11 colors in the color ramp, particular data points are indicated within the map to facilitate the overall data representation.

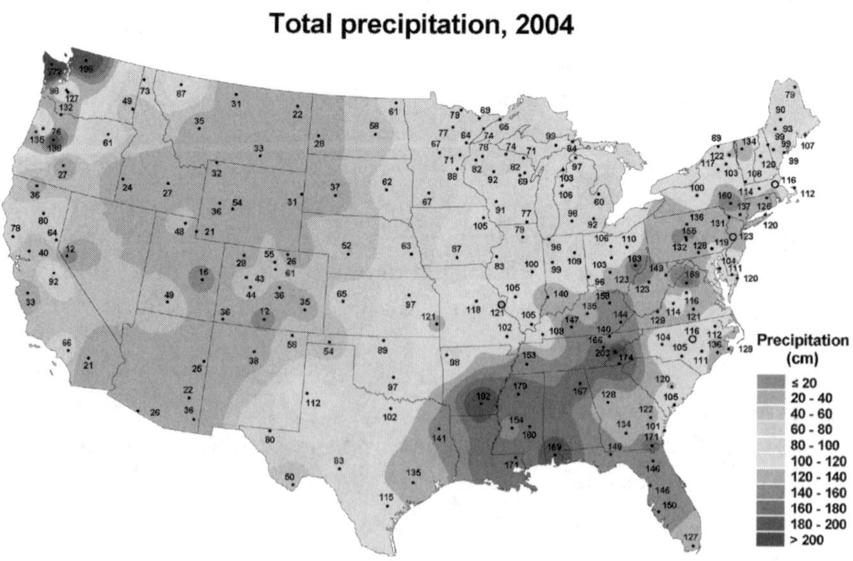

Figure 11.6 An isopleth map of the total U.S. precipitation in 2004. (*Source:* National Atmospheric Deposition Program (NRSP-3), 2005. Reprinted with permission.)

The NADP's isopleth map is used by various professionals to understand not only national precipitation trends but also where precipitation could pose problems. Suppose you were a manager for a major national lumber company and were responsible for picking the location of a new facility. One specification for the facility is that it would need five acres of open land for drying and storing wood. NADP's isopleth map would certainly help minimize the potential effect of rainfall and moisture on the open-air wood storage, and, ultimately, help distinguish good facility locations from poor ones.

Aside from the professional realm, isopleth maps are commonplace in television and published weather reports. Temperature belts and pressure bands are typically illustrated with isopleth maps and high-contrast color ramps (i.e., red means hot and blue means cold). Just imagine how much harder it would be to understand these weather reports if broadcasters used choropleth maps or, even worse, graduated symbol maps. The successful use of isopleths, as with the other map types, is defined by the data being displayed, as well as the anticipated use of the map.

Mapping Data, Classification, Portrayal, and Distinction

Thematic maps, by design, fortify attribute data dissemination and promote analysis not readily obtainable through data tables. These theme-based representations offer highly useful information only if the information being presented is organized logically. Effective thematic mapping involves understanding the data to be presented, a sensible breakdown of that data, selection of a proper map appearance, and well-defined map features. Let us briefly discuss these important thematic mapping elements.

Counts, Rates, and Densities

As discussed, each type of thematic map has its own characteristics as they apply to the data representation. Thematic data can be quantified and represented as numeric counts, rates, and densities. *Counts* are real, whole numbers that represent distinct data entries. Examples of count data include population, number of facilities, and total soil samples. Individual counts are typically represented by a symbol for each item counted. If the counts are too large or exact locations are indeterminable, then one symbol can represent a group of count data (e.g., "one dot represents 100 people").

Rates are a proportion between two counts and are devised by dividing one count by another. Rates are mapped to illustrate how an attribute has changed over time for a specific region or to show the relationship between two attributes. Rate data includes percent of the total and percent change. Choropleth maps are primarily used for rate data mapping.

Densities are similar to rates in that they show a relationship between two counts; however, densities are intimately related to the geographic region. One count in a density is an area of the geographic region where the second or other count was conducted. Densities include persons per square mile and concentrations per square foot. As with rates, a choropleth map is the superior mapping method for density data.

Data Classification

Data classification can determine whether a thematic map is effective. Data classes were briefly mentioned during the map type discussions, though we must now examine the different methods of determining proper data classes. There are four primary methods for data classification: equal interval, quantile, standard deviation, and natural breaks.

The *equal interval* method is self-implied: It breaks the data into classes of equally distributed intervals. For instance, town population data can be broken into equal intervals of 1,000 people, such as class 1 equals 0 to 1,000, class 2 equals 1,001 to 2,000, and so on. Equal intervals offer an unbiased yet simple way to break represented data.

The *quantile* method divides data into classes having equal amounts of features. For example, if a thematic map had 100 attribute features, you could use five data classes with each representing 20 features. Data are ordered by rank in the classes and, as such, the quantile method proves excellent for maps with ranked values. However, one potential disadvantage to the quantile classification is that it distorts the natural distribution of attribute data and may skew the overall attribute data analysis.

Unlike other data classification methods, there is a defined nomenclature associated with quantile maps. For instance, maps with three classes are called *tertile,* maps with four classes are called *quartile,* maps with five classes are called *quintile,* and so on. Additionally, due to the same number of features per class, quantile classifications often produce visually pleasant maps.

Standard deviation classifications create classes that represent standard deviations from the average (mean) attribute value. Typically, the standard deviation method requires an even number of data classes so that there are an equal number of data classes above and below the mean attribute value. This data classification offers simple comparisons between above-mean and below-mean data. One disadvantage to this classification is that a level of statistical understanding is needed to fully interpret the resultant map.

Finally, the *natural breaks* method recognizes that the breaks in distribution may be directly related to the characteristics of the theme being mapped.

GIS mapping software is able to search for these natural breaks in the charting or histograms of the data using a statistical equation called Jenks method. The natural breaks classification maintains uniformity throughout the entire thematic map and avoids any obviously saturated data classes. For example, a choropleth map with 49 out of the 50 United States in one color offers no real analytic advantage. This method of data classification would find minute differentiations in the data and adjust the data classes accordingly, thus producing a more useful analytical map.

Projection and Scale

The topics of projection and scale have been discussed throughout this book because they are, among other things, functional map essentials. Thematic mapping requires the designer to judiciously select elements and features that promote understanding and efficacy. To do this, map designers utilize both map projections and scale to present the data in the best light, so to speak. As we know, map projections are selected to best highlight the region being mapped through minimizing unwanted distortion. Each projection type has variations that may or may not affect the mapped region. Careful consideration must be applied when selecting the best projection, based on an understanding of the desired properties you want to exhibit.

Working in tandem with a projection is map scale. Depending upon the level of attribute data refinement, map scale may be manipulated to best depict the available data. For instance, a total population count of each U.S. state would be best scaled on a nationwide level, rather than state or county level. However, if population data were available for Michigan's counties, a state-level map of Michigan may be warranted if applicable to the user's mapping objectives.

Both scale and projection are wholly user defined and, if used right, can be significant in transforming a data set into a highly useable geographic information product.

Legends, Color, and Symbols

A legend is a map reference that lists and distinguishes various map colors, lines, shapes, patterns, symbols, and annotation. Often, the legend highlights other map-related information such as the scale, source, and projection. Symbols defined in a map legend are assigned to particular data attributes or themes within the map. Map colors, on the other hand, distinguish between similar legend elements and symbols (e.g., colors can distinguish different soil characteristics). Legends, colors, and symbols, when used together, give the user total disclosure of the relevant information needed to discern spatial trends within the data set.

An interesting and important aspect of these mapping elements is that they compose the geographic product's metadata. Back in Chapter 4, metadata was discussed at length. The map legend elements, such as map scale, title, and classification, make up the metadata used to catalog the geographic product within a GIS or other database management system. The more information provided in the map's legend, the greater the document's access and searchable data.

Figure 11.7 offers a sample thematic map legend from a wetlands inventory. Notice that there are very distinct elements within the wetlands inventory legend that define distinct forms of data within the map. We have the map title and/or identifier that label the map for the user and distinguishes the map's purpose. Under "Area Transportation" there are different symbols used for various transportation-type elements. For instance, a thick multishaded line is used for an interstate highway, solid colored lines are used for roadway features, and roadway sign symbols are used for roadway identification. You should also recognize how different colors are used to differentiate between the various features (i.e., state road, streams, and so on). Under "Wetlands" in the legend, color fills distinguish attributes among the various wetland types.

Figure 11.7 also highlights other important map elements such as the county identifier, map scale, map number in the series, and date of creation. These tidbits of information add to the user's scope of understanding and enable universal use.

In short, the sole purpose for legends, color, and symbols is to enable anyone with any background to pick up a map and understand, at least on a minimal level, what information the map is conveying.

Common Cartographic Elements

Before closing the topic of thematic mapping, we must examine the cartographic elements that frequent and add dimension to maps. Thematic maps have two basic separations: the data frame and the map information frame. The data frame is the actual thematic map with themes overlying a basemap. The map information frame is the area of the information product that contains cartographic elements. Figure 11.8 offers an advanced view of the most common map elements

The elements depicted in Figure 11.8 are typical features that are found alongside or below a thematic map (data frame). Since proper map identification is needed for user comprehension, thematic maps utilize a map title block and a more detailed map identifier. The map's title block typically contains the company responsible for the map, as well as the type of thematic map (e.g., a

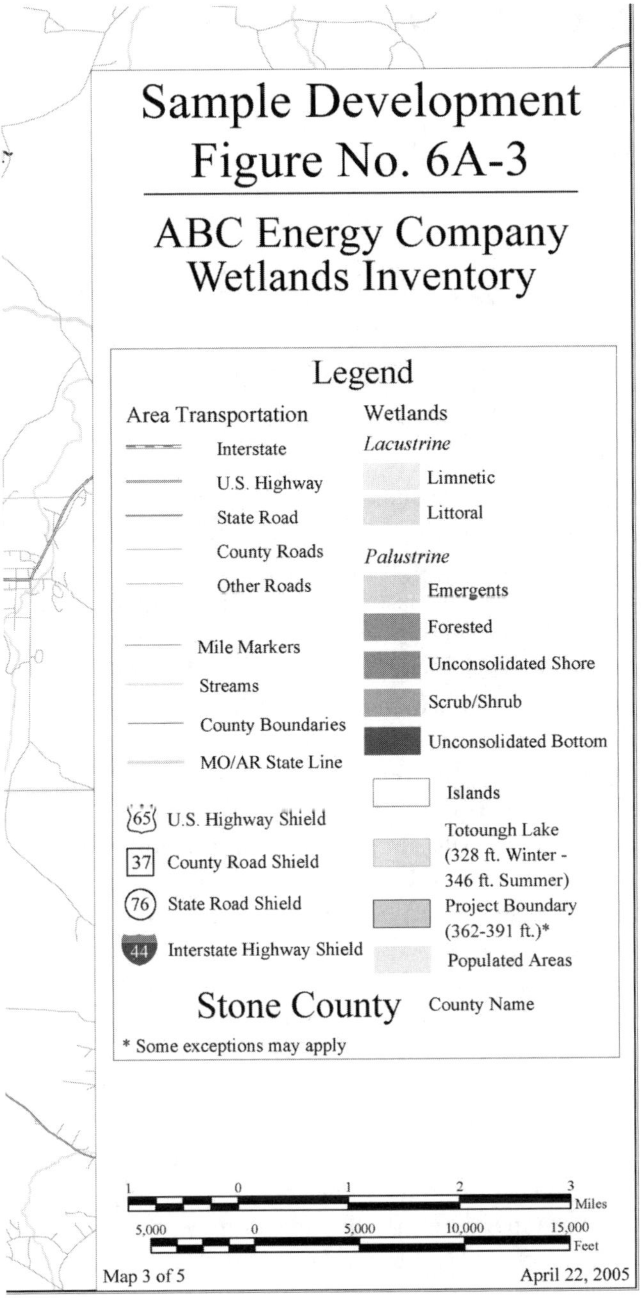

Figure 11.7 A sample thematic legend.

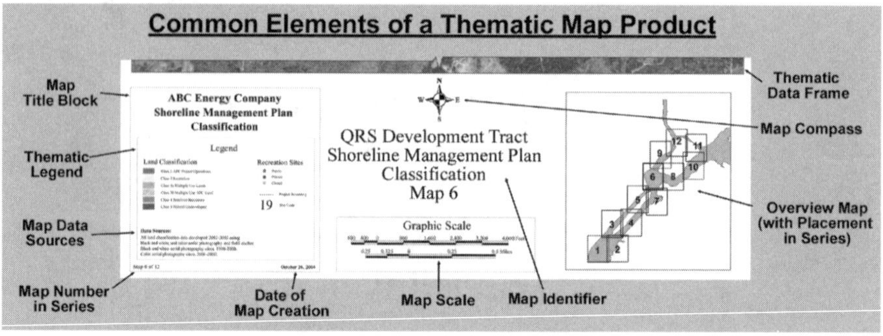

Figure 11.8 Common elements of a thematic map product.

shoreline management plan classification). The map identifier highlights the land involved, the nature of the theme involved, and the map's position (number) within the grouping of area drawings.

As discussed in the previous section, a common element found on a thematic map is a legend. The map legend offers the user a pure "nuts and bolts" understanding of the map's thematic phenomenon. The figure's legend contains an assortment of data class information, with color fills used for land classification and dot/line symbols for recreational sites.

The map's legend may also contain more background information, such as data sources, a textual description of the map's number in the entire series (if applicable), and the official date of creation. This background information will serve as a key component in useful information products.

Any cartographic product utilizes both a map scale and a map compass. The map scale, as previously discussed, helps present the attribute data in as much detail as possible. In fact, scales can be manipulated to portray a larger region of focus, as well as aid in the recognition of actual feature size. The map compass has a simple function: It signifies the true direction of north. Often, the map compass is just an arrow pointing north, while other times (as in Figure 11.8), the map compass is more elaborate.

Another common cartographic element employed is the overview map, which graphically illustrates the individual map's placement among the entire series of maps. For instance, the ABC Energy Company's map was map 6 in a 12-map series. The overview map shows the arrangement of the maps, with map 6 highlighted. As you can imagine, this map element supports user understanding of the region depicted in the data frame and offers insight into how it relates to both the landscape and the other maps.

Overview maps enable individual maps to retain a high level of detail and portray the focus region in an easily viewable format. Just think how ineffective and small featured a thematic map would be if it had to represent all 12 maps in one data frame.

Overall, no matter which cartographic layout and elements are used to describe the attribute data set, the mapping must be consistent, well detailed, and easy to read. These GIS end products are ultimately used by interested people engaged in a solution. It is always important to remember that when the thematic maps are of good quality, the GIS and support efforts are held in high regard. An ineffective or poorly executed thematic map will, subsequently, mar geographic information efforts and vandalize the data credibility of the GIS.

12

GIScience, Engineering, and Related Applications

Up until now we have discussed geographic information systems on a basic level, describing the *what* of GIS with some of the fundamental elements that make the spatial technology tick. Now we must tackle the reasons why a GIS would ever be needed on a project and the advanced, innovative ways GIS is being used today.

As stated earlier, GIS is used in many different industries for a wide range of reasons. This chapter, as well as the next two chapters, attempts to clarify some of the *whys* of GIS with a compendium of different and viable real-life applications.

Using GIS for Technical Applications

In *The Quest for Certainty*, John Dewey, a prominent American philosopher and educator, wrote, "Every great advance in science has issued from a new audacity of imagination." GIS is clearly such an "audacious" technology. Within its short existence, GIS has proved itself an excellent analytical tool and a leading pathway to innovation. GIS's architecture, unlike any enterprise database system, is well designed for the widely diverse implementation field.

Even more dramatic is GIS's usefulness in highly technical and specialized arenas, such as the sciences, engineering, and technology. With these and other technical applications, GIS has helped companies stay one step ahead of competition, while emblazoning their technological contributions and advances. Many of the world's largest and most successful technological companies and agencies use GIS as a means to centralize their information, automate their analytical processes, and analyze the current and future markets.

In tandem with the scientific and engineering advances, GIS helps companies enhance their business operations from financial endeavors to marketing and potential use analysis. GIS's inherited automation offers the opportunity for higher efficiency over time while keeping procedures consistent and information available. The following discussions cover general science and engineering applications, while offering some perspective on the business end of the technology.

GIScience: Principles and Applications

Geographic information science (known as GIScience) is the study of geographic concepts, such as places, things, and phenomena on the Earth's surface, using GISs. GIScience is concerned not only with where objects are on the Earth's surface but also what these objects are and their relationship to other entities. Through GIScience, GIS becomes both an information technology and an Earth science. With its innovative software technologies, GIS showcases information about the Earth's surface (and other celestial bodies) with geographic data and, in turn, offers the visually gratifying form of geographic features. These features supply the specific knowledge about where things are in the world (their locations) as well as what they are (their attributes).

GIScience also provides nearly effortless analytic methods for describing spatial autocorrelation, or the spatial relationships between geographic features. Geographic data are diverse in character. Some data sets are very detailed and contain volumes of attribute information, while others are restricted and modest in size. All of these elements define the broad interpretation of GIScience.

Though GIS is most commonly classified as a computer or physical science, GIScience opens the gates for the infusion of sociology, the empirical social sciences, and history. GIScience is now used for geoarcheology, psychological analysis, social services, and historical analysis. Additionally, this trend of interdisciplinary applicability is most closely echoed by the field of geostatistics, a discipline offering a synthesis between two disciplines: traditional statistics and geospatial interfaces.

Make no mistake, the definition of GIScience is very broad and the proverbial ground it covers will grow exponentially as object-oriented geospatial functions and techniques become ever more important.

Case Study: Using GIS to Study Relationships Between Bedrock Geology and Subtle Topography

GIS plays a vital role in the geosciences and is increasingly becoming a source for revolutionary scientific analysis. One such study involved the definitive quantification of relationships between bedrock geology and topography, where the bedrock contained sandstone and shale, and the topography involved a subtle but well-defined surface. The study area was in Oklahoma in the United States and contained geology fractured by area drainage. The researchers used GIS, importing a digitized basemap of Oklahoma's bedrock geology and a topographical map. Unfortunately, topographical maps prove difficult for detailed evaluation of landforms over large areas and small reliefs, or elevation differentials. The researchers decided to use a digital elevation model (DEM) for a more precise analysis of the study area's subtle topography.

Ultimately, by using GIS the researchers were able to distinguish that the study area's topographic variations rely heavily upon the ratio of sand to shale in the bedrock in areas where sand overwhelmingly exceeds shale. Sandstone was found to be thicker in higher relief areas and thinner in lower relief areas. Additionally, the researchers found that there was no real difference in geology bulk density within these variant regions. Without GIS, these relationship findings would not have been made or determined easily. GIS enabled the researchers to easily quantify the subtle differences in local topography within the different bedrock geologic units—an onerous feat using conventional field techniques. The study has afforded future research and quantifiable topographic studies a viable basis for better land use and an enhanced understanding of subtle topographies.

Engineering Applications

As mentioned previously, the field of GIScience is broad in its interpretation and crosses borders with other applicable disciplines, such as the environment, conservation, mathematics, and engineering. More specifically, the physical sciences have long shared common ground with the engineering discipline in such areas as natural resource design, feasibility studies, and environmental impacts from engineering applications.

Similar to GIScience, the engineering industry has maintained its own history of geographic techniques. From the early days to the present, drawn maps and plans have played a huge role throughout the wide stretch of engineering applications. Amongst the civil, structural, environmental, transportation, industrial, aerospace, mechanical, and electrical industries, for instance, engineers regularly use site maps and plans to depict the proposed scope of work, particular designs, and the proper implementation of such designs in a defined geographic space. Old hand-drawn prints and layout designs showcased early geographic mapping techniques perfectly and provided engineers with a planning foundation. These painstaking hand drawings became extinct with computers and computer-aided design programs. Today, engineers are supplementing CAD with the dynamic drawing accessories offered by GIS.

Today's geospatial standards in the engineering industry are actually inherited from the days of plan views in black pen on a Mylar transparency, displayed through an overhead projector in a dimmed conference room. Black pen markings on the site plan can now not only be accurately scaled and drawn using CAD but can also be geospatially defined in real time and geocoded within GIS. The engineering industry has readily embraced this conjoining of CAD with GIS and has realized GIS's positive impact on company bottom lines, project budgets, quality assurance, safety, and the project schedule.

But GIS goes well beyond engineering site plans and geographic products; engineers use GIS for various engineering applications, such as property delineations, conduit/transmission runs, pipeline siting, marine science, surveying, engineering scenario planning, terrain analysis, and traffic accident research. GISs, for example, help utility companies define their electrical substation placement among distribution networks over entire coverage areas. As another example, oil companies use GIS to pinpoint offshore drilling locations, hyperlinking drilling reports and detailing product quantities for these particular geographic locations.

Some of the more effective engineering applications are oil and water pipelines. The pipeline industry's use of GIS is highly progressive and embraces a wide range of disciplines and engineering methods. Typical pipeline projects involve, to some degree, surveying, GPS and remote sensing, pipeline siting, vegetation mapping, wetlands delineation, natural river/waterway channel analysis, parcel delineation, slope analysis, and defining existing pipe locations. In fact, the pipeline industry has moved in the direction of *trenchless mapping,* or full data collection and scenario planning before construction starts. GIS proves exceptionally advantageous in gathering all data, defining construction scenarios, conducting analysis, and detailing the entire project in one effort without breaking ground. GIS-based trenchless mapping helps the pipeline constructors understand the situations they will face while minimizing expensive project changes, enhancing their public outreach programs, and keeping the risk for litigation at bay.

We could review each engineering discipline and highlight literally hundreds of valid, real-life applications where GIS is used and has made a positive difference. There are many available publications and periodicals that solely detail variant and innovative applications. For this discussion we will take a look at one specific application, New Jersey Natural Gas' Transmission Line Parcel Delineation, whereby the critical GIS functions used are representative for many engineering and utility projects.

Case Study: New Jersey Natural Gas Transmission Line Property Delineation

To prepare for law-mandated public and emergency notifications, New Jersey Natural Gas (NJNG), a utility provider in central New Jersey, required an accurate record of existing properties within 700 feet of its underground gas transmission lines. These central gas lines service a massive area that covers Ocean, Middlesex, Passaic, and Morris counties. NJNG quickly realized the absolute complexity of such a large-scale property (parcel) delineation effort. Accordingly, the New Jersey utility turned to a local geospatial and engineering firm, T&M Associates, Inc., to manipulate the existing data and develop the necessary GIS backbone for the project.

The geospatial system was built upon ESRI's ArcGIS 9 environment. Existing digital parcel information was initially obtained from the state and other sources and imported into the system. Original tax Mylar sheets were scanned into a raster image and consistently referenced with existing tax file naming conventions. This scanned raster image was then rectified with New Jersey's high-resolution orthoimagery (from 2002) and digitized. The digitalization process defined lot lines, boundary lines, and roadways and streams. The outer edges of primary building rooflines (or *footprints*) within the project area were also delineated. The lots created from individual tax sheets were saved as a merged geodatabase file, with adjacent sheets edge matched as necessary. A best-fit approach was used to create the overall parcel layer.

Finally, the parcel layer and subsequent feature attribute table was linked with New Jersey's tax list database, called MOD IV. Unique identifiers and quality assurance protocols were used to facilitate the joining of information, preventing data duplication and parcel ownership errors. In the end, the parcels 700 feet to each side of the transmission lines were successfully delineated and a base GIS was developed for future NJNG efforts.

Marketing and Financial Applications

Aside from the purely technical aspects of the engineering and scientific communities resides the ever-important business bottom line. No matter how advanced a company is or how many innovations a firm has brought to the market, bottom line profitability is the true make-or-break for any business venture. To stay successful, business managers must understand the needs of their market while staying abreast of new potentials. These endeavors take resources: people, time, and money.

The keystones to business bottom line profitability and industry success prominently include business marketing and financial stability. To help managers with their market and financial analysis, business managers are reaching out to new technology and industry trends (i.e., Six Sigma quality discussed previously). In recent years, GIS has played a huge role in conducting valid business sustainability analysis while offering managers an amazing tool for understanding business shortcomings.

GIS offers business marketers the valuable capability of real-time market analysis and customer profiling. Customers can be researched through demographics and purchase histories so that marketing can be redirected (if needed) and the precise customer reached. Some firms advertise GIS-based marketing services that are geared for medium to large businesses of any venue. Internal marketing departments have teamed with in-house or outside GIS consultants to target their market campaigns. In the end, this business marketing strategy has a complementary effect on the fiscal end of the business.

Business financial departments are using GIS to model sales patterns and market histories while aiding in the discovery of profit model inefficiencies. GIS can help accountants prepare or conduct company audits, as well as handle the overwhelming requirements of state- or country-imposed fiduciary regulations. Sometimes, GIS is used for combined marketing and financial analysis when product areas or markets have drastically decreased in sales or new services have proved unprofitable. Commonly, marketing and finance go hand in hand to keep businesses profitable and in touch with the markets. The following case study melds both business ends together through the investigation and interpretation of a rapid chain store closure.

Case Study: Analyzing the Rapid Closure of a Cincinnati-Based Office Depot

In mid-2001, Office Depot, a major chain store for office supplies and equipment, opened a new branch in an eastern suburb of Cincinnati, Illinois. After just six months of operation, the new store closed. Was it a failure in marketing? Was competition surrounding the store too harsh? How could a formidably successful chain have such a fiscal miscalculation? These questions prompted a retail marketing study using GIS analysis tools to offer insight as to the reasons for the store closure.

The branch in question (Beechmont Avenue) was a sister store to five other Office Depots and had competition from five Office Max stores and nine Staples stores in the greater Cincinnati area. The researchers decided to focus on geographic location, since location impacts sales potential and the closure of the Beechmont Avenue branch was related to its low sales potential. The researchers developed a sales potential model within GIS based upon a retail industry standard (D. L. Huff's sales potential model) to evaluate the store's overall performance.

GIS analysis constituted the known spatial distribution and spending habits of potential customers in coincidence with the store's overall highway accessibility, square footage size, approximate travel time, and distance. These characteristics were diagnosed against preopening and postopening data from the Beechmont store, the other greater Cincinnati–based Office Depots, and the area competitors. Through GIS analytical tools and the Huff-based sales model, the researchers were able to ascertain that one reason the store was underperforming was poor customer accessibility. Closing the branch may have been more cost effective than keeping the failing store open, although as noted in the study, if this type of GIS-based analysis had been conducted before opening the branch, management would have saved a great deal of time and money.

13

GIS and the Environment

Since the world's conscience about the Earth's resources began to peak in the mid-1970s, there has been increasing public concern and interest in environmental issues. Lead-based paint, asbestos-containing materials, PCBs, mercury, underground storage tanks, and chlorofluorocarbons (CFCs) have been just a few of the many environmental issues holding public scrutiny throughout the years. However, environmental concerns migrate well beyond hazardous materials and manufactured contaminants. In fact, a cavalcade of public and governmental issues have addressed environmental conservation and those arenas governed by Mother Nature. Animal habitat preservation, deforestation, endangered species, geology, floodplain analysis, water resources, and mining are just a handful of the myriad areas where the public, environmental agencies, and nature convene.

With this ever-growing and substantial public inquiry, state and governmental agencies need accurate information at their immediate disposal that can be quickly discriminated, analyzed, and conveyed to the mass public. To this extent, GIS has become the tool for both the field scientist and the desktop environmentalist.

Using GIS for Environment Applications

Environmental issues became vividly clear to the public in August 1978 when the small, populated community near Niagara Falls, New York, filled headlines. The community was called Love Canal. Hidden underneath this 15-acre neighborhood was a former chemical landfill. After reaching maximum capacity in 1953, the landfill was closed and covered under layers of dirt. Buried within the soil were metal drums that, throughout the years, began to rot, causing hazardous chemicals and wastes to leak into the soil. By the mid-1978, the leaking drums became horribly evident when hazardous wastes began percolating to the surface of Love Canal's backyards and cellars. It was later reported (in November) that 200 tons of dioxin, a lethal chemical, and more than 200 chemical compounds were buried in the canal.

In 1978, the Love Canal neighborhood included approximately 800 single-family homes and 240 low-income apartments. Three schools also graced the Love Canal parcel, of which the 99th Street Elementary School was built almost directly above the former landfill's epicenter. A posttoxic release study revealed a high rate of birth defects and miscarriages among Love Canal families throughout the years and offered documented claims of birth deformities and pregnancy-related problems within the community. Ultimately, the Love Canal incident remains one of America's worst environmental disasters and one of America's most remembered tragedies.

Beginning in the 1990s, GIS has grown as a chief facilitator for major environmental efforts. This technical evolution was inevitable since most environmental efforts are deeply entrenched within the fabric of the land and hold geographic importance. GIS is used for hazardous material issues and natural occurrences alike and has now become the popular instrument for land disputes and owner delineation lines. Researchers are now applying the tool to past contamination for further analysis and understanding. In fact, the U.S. Environmental Protection Agency (USEPA) has reprised analysis on the Love Canal incident using GIS so that environmentalists, agencies, scientists, the public, and the victims have a heightened clarity as to what really happened and a sharpened view of every angle of which people were affected.

The following sections detail common GIS applications as they relate to environmental issues and conservation, though in theory GIS can be used for analysis in many environmental situations. Although cost and feasibility are typically the prohibitive factors for GIS in many environmental decisions, when GIS is applied, the results are sound, intrinsically useable, and many times astounding.

Managing Natural Resources

Natural resources are said to be the Earth's wondrous gifts to humanity. The soil upon which we walk, the vegetation we utilize, the animals with whom we coexist, and the water we drink are all innate elements to life and the recognized gifts of our planet. Natural resources can be considered the prize possession of all humans, although, throughout the ages, humans have neglected to care for these gifts.

There is not a day that goes by without our hearing or seeing examples of human neglect, such as the greenhouse effect, air pollution, toxic emissions, hazardous waste, oil spills, corroding shorelines, and endangered species. During the past two decades, a new surge of environmental conservation of the Earth's resources has surfaced and caught the eye of the public. The management of these natural resources has become the breeding ground for conservationists, public outcry, activism, and lawsuits. This surge has evoked many ideological and creative uses for GIS.

Natural resources encompass rivers and streams, lakes and inlets, landforms, geology, forests and woodlands, mines, wildlife habitats, and ecosystems. The management of these natural earthly gifts involves a deep understanding of each system and a proactive plan for these resources to remain unharmed. Natural resource management covers a large platform for potential analysis, from reclaiming brownfields and facilitating wildlife habitat protection to defining oil spills and delineating land classification. GIS is used by conservationists of all types to define potential infringements upon the resources from nature, as well as any residential, commercial, or industrial imposers. It helps form a stable cadre of geographic information and analysis from which to solidify environmental campaigns or to cast the proverbial stone. Conservationists look toward GIS for the position they need in the fight against combatant corporations, developers, and residents.

But GIS doesn't satisfy just the conservationists; it holds valid analytical and environmental defense support for those on the other end of the conservation fight (i.e., the corporations, developers, and residents). For instance, GIS may be used by a developer to show the self-benefiting, attractive results of little or no intrusion against the natural aesthetics of a lake. Dam owners may use GIS to show how a dam removal would positively affect the area, while not disclosing the owner's high-yielding advantages. Whatever side you are on, GIS proves to be a magnificent propaganda tool.

Many times GIS is used toward natural resources when there is no head-to-head fighting at all, but just as a way to quell rising public concern or help with preemptive actions as to appropriate for the specific natural resources. Energy companies tend to use GIS to explore the impacts of operational by-products, such as the impact of rising waters on wildlife, and uses analysis from the GIS products to determine if actions are warranted.

Case Study: Lebanon Natural Resources Management with Participatory GIS

Entrenched within the Mediterranean, Lebanon boasts a diverse socioeconomic society throughout its relatively small population (4.1 million people). Remarkably, Lebanon is a war-torn land. By 1991, 17 years of war ended, leaving Lebanon overwhelmingly disabled. Its once profitable agricultural industry was deteriorating, and the country looked toward its natural resources for help. However, after years of brutal war, Lebanon's resources were unmanaged and in disrepair.

The Environmental and Sustainable Development Unit (ESDU) at the American University in Beirut took notice and began a series of research programs that meshed technology with community support. The ESDU developed the "participatory GIS" or PGIS, whereby the community provided participatory research, land data, and information for a GIS. The program approach proved highly successful, evidenced by the fact that the ESDU utilized PGIS from 1995 to 2003.

A participatory GIS is a fresh approach to getting relatively accurate measurements and geodata for areas that are hard to measure. The nature of participatory GIS offers many incentives to native volunteers to measure and register land-based and geological features. Such incentives include comprehensive area resource identification and profitable agricultural analysis. In exchange for these incentives is the ability for a highly customized and unique data set. Participatory GISs are gaining popularity in underprivileged communities worldwide and within the GIS community.

Lebanon used PGIS for various natural resource management applications, such as to delineate indigenous agro-ecological zones, plan orchard development, delineate poorly or inappropriately used land, plan rainwater harvesting reservoirs, and develop georeferenced databases for the study area.

Ultimately, PGIS served Lebanon well, providing accurate georeferenced land information that has proved invaluable to researchers, developers, communities, and land stakeholders alike. The participatory GIS information products have given Lebanon a new lease on their natural resources and a feasible approach for socioeconomic growth.

Case Study: Community-Based Natural Resource Management Program in Namibia

In Namibia, Africa, GISs are playing an important role in supporting community decisions as they relate to the management and use of Namibia's natural resources. Namibia's communities are using GIS information products, particularly printed geographic maps, to manage the natural resources on an individual community level. Namibia's Community-Based Natural Resource Management (CBNRM) program involves participants in every community to encourage the proliferation of sustainable development. Namibia's community-based program is very similar to Lebanon's participatory GIS program, but with a heavier focus on natural resources.

The CBNRM's use of geographic data and information is distinct, primarily land-use planning, resource conservation, game and wildlife monitoring, tourist and commercial development, and communication. One of the chief activities threaded throughout the various GIS uses is boundary delineation, especially for conservancy and land development.

Namibia's community GIS program relies heavily on an intercommunity sharing of geographic data, information products, and tools. The program runs on a needs-based approach and involves dedicated community field users who trust the accuracy of the GIS information, have fostered a set process to delineate community needs, and have built a reservoir of information to ensure the program's longevity.

Environmental Investigation and Remediation Applications

A continuously growing area for intensive GIS use is in the hazardous waste arena. With the uprising in brownfield redevelopment, environmental impact statements, building toxics, and hazardous material remediation, GIS has become the premier project database, saving companies and agencies substantial amounts of money over long-term projects.

Interestingly enough, the USEPA is the largest user of geospatial information as it relates to the environment. The USEPA Web site, http://www.epa.gov, boasts a fully hyperlinked and geospatially coded (or *geocoded*) mapping tool, whereby any user can search a U.S. location and retrieve graphically enhanced maps depicting all recorded environmental concerns. The Web site provides a wonderful example of GIS use in an environmental application.

Hazardous materials come in many forms and can infiltrate the human ecosystem in an innumerable number of ways. Just a sampling of the possible environmental episodes include leaking underground storage tanks; toxic releases; the presence of volatile organic compounds (VOCs) and semi-volatile organic compounds (SVOCs) in soils; and micro-fiber silica and polycyclic aromatic hydrocarbons (PAHs) in the air. As with the USEPA site, GIS can and is being used to geographically reference release locations for various agencies and private firms.

GIS has made an indelible mark on the way sites are remediated and monitored. In particular, heavily contaminated sites such as the USEPA superfund sites and chemical dump sites use GIS to track soil boring locations, with hyperlinked sampling results. For environmental site assessments, GIS is used to organize Sanborn maps (historical site characteristic maps) that detail site history, to depict locations of potential contamination, to hyperlink sample analysis, and to represent underground structures, such as storage tanks, and known soil contamination.

On asbestos and lead-based paint contaminated sites, GIS is sometimes used for large contaminated buildings and facilities, whereby asbestos "hot spots" and known lead-based paint locations are meticulously defined in the system. Sample analysis, air-monitoring results, abatement information, and containment locations are delineated, and sample chain-of-custody information is accurately tracked.

Case Study: Illinois Underground Storage Tank Monitoring and Inspection Program

Within the United States, Illinois has more than 28,000 facilities with underground storage tanks (USTs). These facilities range from industrial to commercial uses, though many are simply gas stations. USTs hold petroleum or other forms of fuel in metal-encased tanks. These systems are well built, but after years of neglect and metal shell deterioration, fuel can leak from the storage tanks into the soil. This causes severe human health issues. Every state in the United States requires owners to report USTs.

Illinois needed a comprehensive yet efficient way to manage the locations and information associated with its USTs statewide. Many former gas station sites still have USTs below ground with newer building structures above. These USTs are of greatest concern since they may not have been drained and closed properly, or residual fuel may have leaked into the soil during tank closure. Fuel-laden soil creates VOCs and SVOCs that are known carcinogens. To help accurately monitor and inspect USTs throughout the state, Illinois's Office of the State Fire Marshal (OSFM) turned to GIS.

The use of GIS virtually reinvented the way OSFM handled USTs, from the everyday monitoring and inspection to the data retrieval and data quality. Before GIS, the field inspectors carried boxes of files with them and had to complete nearly 20 different forms to collect the necessary tank inspection information. They then had to mail these forms to OSFM headquarters for processing and entry into the database. Now that GIS has been implemented, the field inspectors need only their laptop computer and GPS receiver. Tank information is entered directly into the computer and the once-tedious forms are quickly completed. Sites get geocoded with the collected information, forms, and GPS location data. Field inspectors can create maps to locate tanks and facilities, as well as perform spatial queries on facilities. In the end, GIS has made Illinois's UST program highly efficient, flexible, and accurate.

Case Study: The USEPA's Disaster Dioxin Air-Monitoring GIS System

After the September 11 terrorist attacks on the World Trade Center in New York City, the world became horribly familiar with air carcinogens and how latent environmental hazards can become time bombs to human health. With the Trade Center disaster, large, concentrated amounts of dust were propelled throughout the Lower Manhattan area, reaching upper Manhattan and parts of Brooklyn, Queens, and Staten Island. Within the fine dust were varying levels of carcinogens such as asbestos and silica fibers. In Lower Manhattan there were many other elements of potential concern, such as fiberglass, mercury, PAHs, PCBs, lead dust, dioxins, and trace metals. The collapse of the building structures caused the material to granulate and become extremely fine dust, an ingestible state that proves most dangerous.

The USEPA used GIS to assist in the analysis of the laboratory results and the field samples. The USEPA set up fixed air-monitoring stations throughout New York City, with a concentration of stations in Lower Manhattan. These stations produced results that were fed into the GIS, from which environmental exposure trends were determined. The GIS information products were used to disseminate accurate, up-to-the-moment results to the public through USEPA's Web site and media presentations.

The GIS products served the numerous environmental agencies and scientists with an accurate geodatabase of sampling results that were at their proverbial fingertips, while offering the nonenvironmentalists a visual depiction of the concentrations and areas where environmental exposure was greatest. The USEPA response was a marvel in the way geographic information and environmental analysis can collaborate to serve a wide range of applications for a much greater array of users.

Shoreline Management Systems

GIS use with shoreline management has provided owners and conservationists a stable repository of geographic landform data and a tool to provide environmental impact analysis. A shoreline management system is a proactive structure by which a shoreline is planned, managed, and/or delineated, with potential human and natural encroachers easily determined. Coastal planning, shoreline planning, and wetlands management are popular topics for a wide range of individuals, including developers, environmentalists, regulators, and lawyers. All three shoreline systems are linked in scope and audience, as well as through their proclivity for dispute and participant division. GIS helps all participants understand the overall scope of the effort in dispute and offers GIS information products (such as maps and land classification charts) to help end these disputes through information.

Environmental regulators use GIS to legally validate their shoreline and coastal management planning efforts, while conservationists use GIS to aid their efforts to preserve natural areas. Developers typically use GIS to further communicate their project's scope to residents and community groups while facilitating the evaluation of the potential area impacts.

GIS information products result from careful shoreline management planning efforts that can lend tremendous credence to any environmental management undertaking and potentially influence political opinions or legal decisions. You might be asking yourself, "How does GIS really do this?" GIS increases the visibility of the effort and showcases facts to substantiate claims, ideas, or complaints. The ability of GIS to utilize plentiful, accurate, and freely available geographic data can be the "equalizer" in big business or bureaucratic environmental litigation. Any users with modest GIS knowledge and desktop computer proficiency can build a system to "make a case" for their shoreline, coastal, or wetlands management cause.

In the United States alone, shoreline, wetland, and coastal management planning is truly big business. Management planning is backed by the government through enactments to state and federal law, including the USEPA's Coastal Protection and Management Act of 1995 and Coastal Protection and Management Regulation 2003, the U.S. Geologic Survey's Shoreline Management Act of 1971, and the USEPA's Wetlands and Watershed Act of 1997.

Case Study: The Coosa/Warrior Shoreline Management GIS—Alabama Power Co.

As a direct result of the U.S. Federal Power Act of 1935, the Federal Energy Regulatory Commission (FERC) was granted the right to regulate nonfederal hydroelectric projects. Owners of hydroelectric facilities must periodically relicense their facilities with the government in order to reap governmental aid and receive revenue from their hydroelectric enterprise. These relicensing efforts are time consuming and costly and must meet FERC and local agency requirements. Requirements are strict, cover various areas of importance, and usually involve the management and planning of impacted shorelines and owner-occupied coasts.

As part of Alabama Power Company's relicensing efforts, a geospatial database was developed to help facilitate the relicensing of seven major hydroelectric facilities on the Coosa and Warrior watersheds in Alabama and Georgia. The project involved spatial database design, spatial analysis, image processing, environmental shoreline management, and land classification.

A complete Shoreline Management Plan (SMP) system was developed for Alabama Power's work at the sites, classifying more than 3,000 miles of shoreline and project areas. With the application of GIS, the SMP mapping effort consisted of more than 150 map layouts that were developed for public presentation and project team analysis.

Other Environmental Applications

The list of geospatial environmental applications is far more extensive than the popular areas discussed in this chapter. You can say that the list is virtually ever growing. Through user creativity and out-of-the-box thinking, GIS has reached the small corners of the environmental industry and instilled renewal into the traditional project management controls. Areas such as pollution prevention, groundwater modeling, environmental compliance, and water resources (among others) have gained new life through GIS and as a result have promulgated higher efficiency rates for technical analysis and the dissemination of information.

By its very nature, GIS is fully migratory within the industries and, as such, has shifted into firm practice worldwide. Whether because of its power of efficiency or its sheer management capabilities, GIS has proved itself an incomparable project tool. Realizing this, many large consulting firms and organizations now have in-house GIS capability and thus offer geospatially based solutions as part of their core environmental offerings.

14

GIS and the Public

Since the advent of the Internet, more and more people, whether they know it or not, use GISs on a frequent basis. Although the Internet and GIS are separate systems, some Internet sites offer GIS-processed information, such as turn-by-turn driving directions or up-to-the-moment weather information. The public has the ability to search neighborhoods for health and safety hazards, search communities for public notifications and alerts, or even find out the air quality index for that specific day in any defined location. The general public relies on these sites for free, quick information, though the users rarely understand that the information they are manipulating is supplied from a GIS or that they comprise the majority of Internet-supported GIS users.

The general public is also the largest group uniquely affected by GIS technology. Information systems have infiltrated everyday life, from tangible products, such as news and home electricity, to abstract products, such as emergency 911 system routing and port vessel tracking. The reality is that GIS and the public are intertwined, and as the technology grows in use, the number of ways one person is affected by GIS becomes incalculable.

A Public Need for Information

Without question, it is vital that the general public have a source or group of sources that provide accurate information. A means to easily distribute geographic information to the public can sometimes mean life and death. For instance, during the September 11 attacks in New York City, the NYC Police Department used GIS to maintain public safety, control Lower Manhattan access, and disseminate community information. Through various state and federal Web sites, the public had the comprehensive ability to access area maps, timetables for redirected traffic patterns, rescheduled public transportation, information about community meetings, and results of air quality samples. This intrinsic use of GIS helped maintain control of a very unstable situation while keeping the public well informed. During the emergency, the public demanded accurate information quickly and continuously, and a GIS proved to be the most comprehensive and easily implemented of solutions.

This need goes much further than just community safety, disaster, and emergency situations; the need resonates throughout public schools systems, public transportation systems, and within the policies and regulations that govern facilitation. To this end, GIS has infiltrated the populace as a subtle but powerful information medium. It can be suggested that the immediacy of GIS information products, in tandem with an Internet delivery, makes daily newspapers and the evening news seem like "pony express" reporting.

Nevertheless, immediacy is not always the main propellant for GIS use. Many times the public user requires the power of the information database rather than data immediacy, since it offers immense public documentation catered to a specific application. For instance, school principals can monitor crime statistics in a school's district, and social service providers can tap into local demographics and social structures. These types of public applications necessitate accurate, time-nurtured data rather than up-to-the-moment data.

Similar to the other disciplines of GIS (i.e., education, business management, environmental, and so forth), public praxis is evolving and is limited only by user creativity. The following sections highlight typical applications for the public's use of GIS and information management. As you will quickly realize, the general public comprises a very diverse suite of users that viewed macroscopically is wide ranging and viewed microscopically is highly user particular.

Managing Public Health and Safety

Public health and safety have always been subjects of debate, interest, and downright concern. From the early days when there was minimal medical knowledge and no technology, people with a communicable disease such as tuberculosis could infect the neighborhoods or towns where they lived. News of a new case would send fear racing throughout the communities. Neighbors would inform others of new cases. People fearful of infection would stay indoors or somehow try to hide from the disease. Invariably, others would be infected. The end limits to any disease management efforts would be to either quarantine and treat the infected or wait for their death.

Nowadays, medical knowledge is painstakingly sophisticated, and the once dreaded communicable diseases are fervently managed by professionals. The public fear of such community-infectious diseases has tapered to mere interest. However, the concern about deadly new outbreaks has not diminished. As in the days of old, public information remains the first and best barrier to a total outbreak, and new technology provides a second barrier for professional analysis and understanding. As you can imagine, GIS plays an integral part within both barriers and the meticulous spatial tracking and management of disease.

Disease management is not the only public health and safety issue where GIS plays a management role. Scientists and medical professionals use GIS to spatially analyze areas for air contaminants and toxins, higher rates of miscarriage and birth defects, spikes in asthma and other respiratory stressors, proficiency in child inoculations, increased worker safety concerns, and occurrence of repetitive motion injuries, just to name a few. GIS aids in public awareness programs, offering accurate information online while providing health and safety analysts a compendium of source data.

The process of managing public health and safety is imperative to lessen human loss or injury when outbreaks and emergencies occur or as an ongoing medical prevention program. GIS is used to inform the public of "hot spots," or places to avoid, while indicating information for preventive measures and locations for proper treatment. Many governmental and state public health agencies use GIS technology to identify or defuse a potential health safety threat, conduct communal human health risk assessments, make better decisions in terms of public safety, or just stay ahead of area health trends. The superfluity of demographics, client data, and resource distribution is among the greatest reasons why a GIS is used.

Case Study: Detailing Areas at High Risk for Elderly Fall Injuries to Enhance Injury Prevention Programs in Alberta, Canada

Researchers from the University of Alberta's Division of Emergency Medicine, the Health Surveillance Branch of Alberta Health and Wellness, and the British Columbia Rural and Remote Health Research Institute conducted an investigation detailing the geography of fall injuries in the elderly within Canada's Capital Health Region in Alberta. The researchers used GIS to analyze the spatial character of fall injuries within the study area in an effort to enhance fall prevention methodologies. The data set for this particular study comprised Alberta residents greater than 66 years of age who visited an emergency department for a fall. Postal codes from emergency vehicle and department reports comprised valid location data.

Using GIS, spatial analysis was completed to identify general fall injury patterns and high-incidence areas in both Alberta proper (such as rural and medium-population areas) and within a domicile (such as on a stair or off a ladder). The researchers relied heavily on geographic information products, such as empirical Bayes estimate maps of incidence data. Empirical Bayes estimates invoke British mathematician Thomas Bayes's theory of statistical sampling, which states that a probability distribution can be assigned to an unknown parameter in a statistical problem. In this application, the researchers assigned an underlying probability distribution to the fall incidence ratio. In the end, the researchers proved their argument: that fall injuries in the elderly occur in distinguishable areas, making region-specific fall prevention programs possible and definitively beneficial.

Disaster: Planning, Response, and Recovery

We have already discussed in some detail how a GIS can be used for information distribution in emergencies or for analysis in the wake of a disaster. However, there are additional functions relevant to natural and human-made disasters that we have not covered, namely disaster planning, response, and recovery.

As a result of the recent natural events and terrorist activities, the need for a heightened focus on disaster planning is sobering. Property owners, bankers, insurance professionals, and state and federal agencies have taken preliminary steps to increasing their disaster preparedness, while being proactive toward disaster response procedures and the minimization of lost business days during recovery. As you might guess, GIS has been playing a greater role in these areas.

Disaster planning involves understanding historical events, analyzing each historical event's "before" and "after," and learning to better deal with a similar situation. Field- and satellite-based GIS data in time-lapse maps are used to study events ranging from the levels of floodwaters and the nature of particular tornado damage to past errors in fire safety and building collapse characteristics. Disaster planning highlights offsets to catastrophic ruin and ways to back up current systems. Overall, disaster planning covers a wide ground, but it can be reigned in through the analytical and database power of GISs.

Disaster response goes hand in hand with planning, but with a deeper focus on past response–related errors. The main purpose of a response is to lessen the amount of loss, be it human loss or material loss. Every past error indicates how a response should not be conducted, while highlighting those areas where additional planning is needed. A well-planned GIS can help route emergency personnel to the correct area or alert them of the correct protocol, transplanting accurate data for potentially fatal guesswork. A GIS can calculate the best route, querying the most recent street-position data, construction detour data, and traffic information. Using a GIS for disaster or emergency responses can be the time-saver needed to save critical material or to save a person's life.

This leads us into disaster recovery, which is the readjusting of a disaster event to a level of normality or stability. For example, banks implement offsite data recovery stations, where vital financial information is transmitted and can be instantly recovered in the event that the origin bank is incapacitated. Disaster recovery goes much further, such as in the reoccupancy of buildings or the desperate search for human life in a building's rubble. GISs are used to inform (once again) rescuers and emergency personnel of vital location data, such as the number and identity of building occupants from an electronic security badge database or maps of predisaster floor layouts. The main purpose for disaster recovery is to minimize loss while stabilizing the situation in a timely manner.

Transportation and Information

Public transportation systems have always been a natural candidate for embracing GIS and the geographic information products the technology offers. Subway maps, bus routes, highways and public access roads, and ferry paths have historically involved graphical street maps overlaid with route paths, route numbers, and legend symbols. Maps offer the general public an easy-to-understand bird's-eye view of the entire transportation system and engage the simple human task of recognition. There are no in-depth text instructions to comprehend or composed tables to siphon through in order to find the desired information. Just graphics, labels, and overlays: the basis for GIS.

With the use of geographic information products, public transportation and the distribution of transportation-related data are much more efficiently facilitated. Transportation GISs not only can map routes for the public user but can also detail construction locations, alternate routes and reroutes, city gridlock areas, and sites of heavy congestion. These systems help keep the public commuter well informed while offering a means of informed transportation choice. This side to GIS use in transportation is extremely visible to the public user.

There is another side to public transportation that is less visible to the end user and much more instrumental in application. GISs are used to maintain subway tracks and roadway, maintain transportation fleets and equipment, identify system faults, route emergency vehicles, pinpoint accidents, and make immediate changes to information products. These applications, of course, are just a sampling of the copious functions that public transportation managers, engineers, and operators require.

Going one step further, transportation planners and engineers use GIS to better analyze particular traffic situations in any study location. These professionals use spatially defined or geocoded data to help discern the causes and effects of the particular situation, while the information gained from this type of geospatial analysis aids in the development of a practicable, cost-effective solution. A GIS enables transportation professionals to build upon existing transportation network data, update information with GPS, and easily integrate spatially referenced attribute data.

You need not look too far from your own town or nearby city to fully observe the power of GIS in relation to public transportation and information dissemination. Visible to the public or not, many of today's smaller public transportation systems (and almost every large public system) are now spatially enabled in some fashion. Check out a transportation Web site for a local agency for indications of geospatial data. If you want to grasp how completely public transportation and GIS can meld, take a look at Portland, Oregon's TriMet interactive system map at http://www.trimet.org.

Case Study: Portland's TriMet Transit Tracker Online

Portland's TriMet system offers its public ridership real-time bus information and train (TriMet's MAX trains) arrival data through an ESRI ArcInfo-backed system called Transit Tracker. The real-time information is accurate wherever the user may be. The system works off GPS mobile tracking, wireless technology, and a GIS. Transmitting sensors on the public buses send location information to the Transit Tracker system database, while sensors built within the train tracks transmit train speed and sensor location. The system displays actual bus locations and calculates an estimated arrival time for MAX trains based upon transmitted information. Public users can access this data wherever they are through handheld devices, the Internet, or telephone.

TriMet's interactive system map uses a high-resolution aerial photograph in conjunction with bus and train data layers. Maps highlight bus stops, MAX stations, points of interest, and park-and-ride locations. As with any GIS map product, map views can be zoomed in and out, information can be retrieved about stops, stations, and routes, and specific locations can be searched and viewed. An interesting feature to TriMet's system is their MAX train "up-to-the-minute arrival time" window that displays train arrival countdowns. Portland's use of an online GIS for public transportation is highly progressive and will set the path for future public transportation information systems.

Homeland Security's Third Eye

Although we have discussed GIS's role with disaster planning, not a day goes by when the phrase "homeland security" is not heard, along with the notion that something more is needed for terrorist events than pure disaster planning. In these undetermined times, the subject of national security is of great importance to both the government and the public. Governments want to maintain a status quo on threat control and homeland sustainability, while the public wants full information disclosure and a stable sense of life safety. Security analysts and managers now use GIS in conjunction with other in-place protocols for threat and vulnerability assessments; security control procedures; sea craft and aircraft tracking systems; surveillance, detection, and intelligence gathering; data sharing; and overall preparedness.

The interesting aspect about using GIS with domestic security is that typical security projects meld the public applications just discussed: public safety, disaster planning, and transportation systems. Its use with public safety is seemingly obvious. GIS offers a means to identify safety concerns and areas for potential vulnerability, whereby officials can close off these vulnerable areas to the general public. GIS products enable agencies to identify and subsequently control public access areas to help maintain public safety and enhance early threat detection.

GIS offers powerful capability in understanding how a safety plan will work in the event of a disaster or terrorist situation. The technology helps security analysts detail the best areas for a rapid mass public exodus from any area, as well as defining the easiest paths for emergency personnel. GIS proves to be most useful with interagency information sharing, in addition to serving as a central repository for information-gathering exercises. This single system, single data methodology may be the deciding factor in preventing a terrorist attack or in maintaining control during a sudden, disastrous event.

The transportation portion of domestic security revolves around mass public threats and evacuation protocols. Because transportation systems handle a large, diverse ridership, with people looking over unknown faces every day, the importance of early detection is critical to human safety. GIS and GPS are used to monitor train and bus locations and can help officials identify any abnormalities with service. These abnormalities can act as the early detector in public safety concerns or hostage and terrorist situations (on September 11, planes had abnormal flight patterns). Additionally, GIS can collect ridership data for any given moment, which can aid in vulnerability analysis and rescue/recovery initiatives.

Traditionally, in mystical circles a person's "third eye" was considered the eye that saw the unseen and invoked a person's extrasensory perception. GIS is quickly becoming the third eye for worldwide homeland security protocols, since it helps analysts "see" what is not readily obvious and helps keep agencies one step ahead of terrorists. In reality, the technology's automation is its greatest asset to these security professionals, while the data's accuracy and consistency enables more concentrated and successful threat mitigation measures.

Public Education Applications

GIS technology has become the newest tool for public education analysis and proliferation. GIS is now used in tandem with teaching curricula to enhance a student's ability to comprehend lessons and perform hands-on problem solving for complex problems. Science, physics, mathematics, Earth studies, and chemistry are among the major subjects most positively affected by GIS, though the scope of in-class analytics is wide. The information products form intriguing, exciting manipulatives that tend to engage students and help progress their understanding of the topic.

Teachers are not the only educators using the information technology; school officials, principals, district leaders, and superintendents have begun using GIS to better understand their district's or school's demographics. Educators and school officials can evaluate their school system occupants through a mix of demographical data that ranges from racial background and economic status to familial situations and standardized test scores. GIS offers a new level of detailed analysis that helps educators and officials pinpoint where deficiencies lie and successes hide. When used correctly, a GIS can be a potent facilitator for a better school system.

The power to conduct accurate educational analysis through information sharing between districts/schools is far more effective than with past methodologies. A GIS, by default, becomes a central clearinghouse for school data, which aids in quantitative and qualitative educational understanding. School officials can precisely rank their schools with other same-level schools, district leaders can compare their successes with those of other districts, and superintendents can monitor the overall effectiveness of their public education domain.

Using GIS in tandem with education helps students and educators become community aware and more firmly involved with the events and activities around the school. This interchange of information breaks the spatial barriers of traditional learning and gives everyone involved a refined vision of data exchange while making each user aware of their community's spatiality.

Aside from the educational advantages of GIS, school facility managers can implement a Web-based GIS for campus/school facility information. Students and educational personnel alike can easily access real-time campus maps, building information, campus activities, and community notifications. Facility managers can use such a system for building inventories and utilities, as well as to dispense facility closure and emergency evacuation information.

In the same vein as facility information, the campus community can use the technology to tap into available neighborhood information for identification of potential threats, to learn about local evacuation procedures, and to realize areas of caution. Some district police departments conduct spatial crime analysis through geocoding community felonies (i.e., turning crime addresses into grid coordinates) to keep neighbors, neighborhood watch groups, and officials informed of local incidents. The availability of neighborhood data depends upon the capabilities and funds of the particular community. Nevertheless, through GIS, public education participants have a viable means of being zealous in their community while encouraging active learning and participation.

Redefining Public Policy

The field of public policy remains one of the most radical arenas for GIS implementation. That stated, one reason for its radical prowess is public policy's diverse and ever-changing atmosphere. Public policy covers every agreeable public domain, spreading from education and the environment to social services and international applications. If that weren't diverse enough, public policy initiatives involve the lawmakers in government, the decision makers in industry, and the eventual public end users. In short, public policy GIS applications cover every niche we have discussed so far in this book.

Since public policy's applicatory enormity can be overwhelming, let me define its common scope for spatiality. Typically, GIS is used to help secure governmental financing, define public issues to the legislators and voters, determine the effectiveness of current policy, help streamline new and existing governing protocols, and identify voids in foreign policy.

Governments finance various efforts, such as agency requests (e.g., war analysis and border patrol), industry innovations (e.g., power generation and aircraft development), and public needs (e.g., public assistance programs and education). Consequently, a deeper understanding of such financial distributions and contributions is required, and the overall effects of these fiscal changes must be adequately assessed.

To simplify fiscal assessment, GIS maps have become the popular tools of public policy applications. Maps can provide straightforward graphical depictions of current policy and highlight the advantageous elements of proposed policy changes. Color scale maps tend to be the presentation of choice for policy makers, since the graphical displays are easy to understand and can be manipulated for any specific purpose. For example, a request for additional fiscal allotment to the U.S.'s Women, Infants, and Children (WIC) program can be proved to be warranted with the help of a needs evaluation map using a confirmed data set and a differentiated color scale. The shades of color can signify increases or ranges in a target need and give a visual, easily distinguishable peek at how an allotment of funds can fill the understood needs.

Going beyond fiscal analysis and the definition of public issues, GIS has been a proponent of public policy education. Many times agencies need to educate the public on their current policies to enhance public awareness of active programs, prevent misunderstandings, and diminish unnecessary outcries for policy reform. GIS can play and has played a huge role in public policy education through detailed maps, explanatory literature, and online information depots. With GIS-enhanced policy education programs in place, the general public is better informed, and ultimately, the policy making process is progressively focused and successful.

Case Study: Public Policy Education in Gaston County, North Carolina

In 1988, the Quality of Natural Resources Commission (QNRC), a committee of appointed citizens in Gaston County, North Carolina, was chartered to identify the county's air, surface water, and groundwater environmental concerns. A substantial portion of QNRC's work was to recommend policy changes or new policy to protect Gaston County's natural resources. Early in the project, the committee was faced with incomplete and fractured data, as well as data too technical for effective policy making. The QNRC's work progress quickly became stifled, at which point a multidisciplinary group from North Carolina State University (NCSU) stepped in to help. The group undertook a two-phase approach to the study, wherein the first phase highlighted a countywide environmental assessment and the second phase involved refining that assessment and developing a public policy education program with educational materials.

Throughout both phases of the work, the NCSU-QNRC team used geographic information systems. In phase one, GIS was used to help the research team analyze the nature and impact of Gaston County's environmental problems. In phase two, GIS was used in a more educational function, with information products being developed and used for policy education tasks. GIS maps highlighting hazardous waste sites, community well locations, water supply intakes, and surface water quality monitoring sites were among the various types of public materials created. These maps were implanted within educational literature and distributed to the public.

The NCSU group developed a public policy education program that piggybacked the publicly distributed educational materials and helped in policy evolution. Geographic information products aided the NCSU group and QNRC to realize the extent of environmental problems within Gaston County, locate the faults within existing public policy, and portray their findings to the public in a wholly comprehensible manner. In the end, NCSU and QNRC met their charter and created an environment whereby Gaston County's natural resources are protected.

15

Getting Started with GIS—An Overview of Environments, Tools, and Data

As we now understand, GIS, by its very nature, caters to visual understanding. A well-planned information product captures the important elements of the location like a snapshot captures moments and delivers it in a digestible format. Individually, text and graphical representation offer essential information for any need, but together, a whole story is told.

The same premise goes for the discovery, demystification, and total comprehension of GIS itself. A textual description of GIS and its many elements such as this book can only deliver a theoretical understanding of the science and systems. The learning of GIS demands interaction and hands-on use for a true and total grasp.

We are now at that crossroad where the essential theory has been discussed and the hands-on portion must commence. This chapter is designed to get you up and running with GIS and to pinpoint key data and software destinations on the Web for you to explore. As you work more with a true GIS environment, the theory will coalesce and an advanced understanding will form.

Getting Yourself Up and Running with GIS

Up until this point you have been armed with GIS theory and are now likely eager to start using a geospatial environment. Like an athlete mastering his sport, the GIS novice must practice in a GIS environment for the theory to fully make sense. But how can you move past the theory if you own no such GIS software and have no data with which to practice?

You can purchase the required software; however, the overall expense may be too great for someone just learning the craft. To gain experience quickly, a more practical solution (at least in these beginning stages) is to try out the free GIS software made available through various outlets in the geospatial and GIS communities. Some of the software offerings have been around for many years, while others are relatively fresh in the field. The array of available programs includes complete GIS environments, component software, basemaps, and geospatial data.

Before moving too far, we will take a brief look at the leading geospatial systems on the market that are considered the industry "standards." Once we examine what is available for purchase, we can set our sights on the easy-to-acquire free programs. Bear in mind that these systems are somewhat scaled-down environments. However, some free GIS environments, such as GRASS (discussed later), are comprehensive and are used as the sole systems in many agencies around the world. With only modest use, you will gain an appreciation of the power and functionality of this available software.

Leading For-Purchase GIS Environments

GIS is a hot commodity in the technical world. Many engineering, environmental, and technology firms use GIS regularly and subscribe a sometimes sizable budget each year for upkeep on the latest and greatest GIS environment. Luckily, today's market is highly competitive and, as such, offers an excellent array of GIS packages from which to choose. As with many software niches, there are many software packages offered, but only a handful are true benchmarks.

The GIS industry has many comparable packages available to the public. Table 15.1 lists diverse GIS environments available for purchase. However, within this listing there are seven GIS environments that lead the pack in terms of package features, flexibility, overall use, and cumulative sales throughout the industries. These systems are ESRI's ArcGIS, Intergraph Corporation's G/Technology, General Electric's Smallworld, Clark Labs' IDRISI Kilimanjaro, Autodesk's GIS Design Server, DeLorme's XMap, and MapInfo.

Although these leading software packages are considered popular and are being discussed within this context, I encourage you to examine the whole range of available GIS environments to identify the right software for your need. All of the software packages on the market are quite similar, but each offers particular variations to define functional autonomy. Equally, every GIS user has a different prerequisite and need, and to select software based on popularity may, in the end, shortchange efforts and minimize project goals. Keeping that in mind, let us take a brief look at these leading geospatial environments.

Table 15.1
For-Purchase GIS Environments

ArcGIS	http://www.esri.com/software/arcgis/index.html
Maptitude	http://www.caliper.com
GIS Design Server and Autodesk Map	http://www.autodesk.com
G/Technology and GeoMedia	http://www.intergraph.com
IDRISI Kilimanjaro	http://www.clarklabs.org
Manifold Release GIS	http://www.manifold.net
MapGraphix GIS	http://www.comgrafix.com/mapgrafix_gis.html
MapInfo	http://www.mapinfo.com
MetaMap GIS	http://www.metamap.com
MyWorld GIS	http://www.myworldgis.org/myworld/
Smallworld	http://www.gesmallworld.com
TatukGIS	http://www.tatukgis.com/Home/home.aspx
TNTmips	http://www.microimages.com/product/tntmips.htm
XMap	http://www.delorme.com/professional/xmap

ESRI's ArcGIS is by far the most popular geospatial environment available. The Environmental Systems Research Institute is an offshoot of the U.S. Geological Survey (USGS) and has grown in popularity since its inception in 1969. Due in part to its ties with USGS, ArcGIS has become the favored system for many U.S. and international agencies. The ArcGIS environment encompasses an integrated set of geospatial products that, grouped together, form a powerful enterprise GIS environment.

ESRI's ArcGIS suite is designed for the private and public sectors alike, making available software for desktops, laptops, tablets, servers, and mobile devices. The GIS software enables users to analyze, manage, and map geographic information and can be deployed throughout an organization by Internet and intranet connections. Users can install plug-ins, called *extensions*, to add functionality to their GIS viewing package. ESRI also offers ArcExplorer, which is a free GIS data viewer that allows basic mapping and spatial querying. Overall, given their solid industry presence, ESRI's GIS suite has remained a strong leader in the GIS software market.

Another strong leader in overall use and popularity is Intergraph's G/Technology and GeoMedia program suites. Intergraph's packages offer powerful analysis, management, and design tools geared toward users in virtually every business sector. Intergraph's G/Technology suite of programs offers specialized industry-specific data models for the various utility, pipeline, water, and communications industries. Intergraph boasts that G/Technology addresses the best practices for cost containment, process improvement, systems integration, customer base retention, and new service opportunities. Additionally, the software was designed as an Open GIS and can subsequently work with many different GIS formats.

Intergraph's GeoMedia software suite provides user flexibility through strong, quality-focused programs designed for efficient map design, presentation, and sharing. GeoMedia allows users the flexibility of desktop, laptop, and enterprise-wide compatibility. Additionally, Intergraph offers GeoMedia Viewer, a free GIS viewer geared toward desktop geospatial viewing and the distribution of data. The combination of G/Technology and the GeoMedia software suites proves a formidable GIS environment.

General Electric's Smallworld suite offers a fresh "feel" to GIS through its advanced spatial technology and seamless existing system integration. The Smallworld suite encompasses a selection of powerful geospatial programs, including Smallworld's Core Spatial Technology, Spatial Intelligence, Enterprise Integration Tools, and Design Manager. Smallworld's program suite is designed for desktop, laptop, and Internet interoperability.

GE's software architecture is somewhat different from ESRI's all-purpose design and Intergraph's general and specialized software programs. Unlike the broad focused programs, Smallworld's design centers upon engineering, scientific, and business-oriented applications. Like its competition, GE's flexible GIS platform is an exceedingly capable geospatial environment.

In an industry with so many competing environments, a good many users prefer user simplicity and image workability over an abundance of functions and project-specific decision-making. For this, Clark Labs offers IDRISI Kilimanjaro, a widely used raster-based GIS and image processing environment. IDRISI was developed as part of an academic research program and has been designed (and redesigned) to stay user friendly and highly accessible, while serving as a benchmark in geospatial standards.

IDRISI Kilimanjaro offers the same advanced GIS and image functions as the other leaders, such as GIS modeling, database querying, spatial data development, and geostatistics; however, the software uses advanced object-oriented development tools that are especially suited for focused research and environmental modeling. Aside from its raster-based functionality and ease of use, its low cost has maintained IDRISI as a viable competitor among the GIS leaders.

No discussion about GIS environments should exclude Autodesk, a strong software company in the engineering and technical fields for many years. Autodesk is best known for its impressive AutoCAD software and has now grown its technical market to include GISs and mapping, since CAD and GIS are complementary. Autodesk Map is the company's GIS environment that is built around AutoCAD and, as such, invokes user comfort through its widely familiar AutoCAD "feel."

Autodesk Map offers the user flexibility in designing, maintaining, and producing maps and geographic data. The software's geospatial capability is solid and acutely comparable to the competition. The program suite supports desktop, laptop, tablet, mobile, and enterprise systems and has a huge following. Through their enterprise GIS system, called GIS Design Server, Autodesk offers easy data integration with reliability and security. Although the user can work on various GIS file types through the software, the most seamless of file exchanges are with files that were originally designed in an AutoCAD environment. That being said, with functionality excellence, user trust, and an already familiar program exchange, Autodesk Map proves itself a GIS forerunner.

Taking the route of minimized expense, maximum function is mapping giant DeLorme's XMap software. XMap is a relatively low-cost, full-featured GIS environment that offers GIS and GPS mapping capabilities. XMap uses a modular design for expandability, interoperability, and enhanced functionality. XMap also supports Open GIS and most GIS data formats. DeLorme's software offers a less expensive alternative for quality, but standard, GIS capability.

This leads us to MapInfo's suite of GIS software, led by their MapInfo Professional program. MapInfo Professional is a popular GIS environment that is comparable to the other leading systems in terms of geospatial capability, analytical tools, and add-ons.

MapInfo's development side is where this software package shines, through MapX, an Active X component that enables alternative software embedding. What this means is that developers can embed mapping applications within other applications, such as Microsoft Word and Excel, Lotus 123, and others. This specific capability added to the suite of other GIS analytic capabilities has made MapInfo highly attractive to business developers and users. Incidentally, MapInfo's product uniquely exemplifies how geospatial environments differ, each offering advantageous characteristics amenable to particular goals.

Now that we have examined the most popular for-purchase environments, let me just reiterate that for-purchase systems require available funding and, as with any purchase, should not be chosen by text alone. You would never dare purchase a new car without taking it out for a test drive. Identify the software packages you are interested in and examine functionality, cost, and file interoperability. Then take your GIS test drive by downloading demos of the software and discovering which software meets your needs and which fall flat.

GRASS and Other Free GIS Environments

To many GIS novices, the thought of spending hundreds, even thousands, of dollars on GIS software is alarming. What may be even more disturbing is the fact that comprehensive software is ultimately needed to provide the user with the full functions, science, and geospatial mechanisms of GIS. As mentioned previously, the proper learning of GIS requires theory intermixed with hands-on use. Without access to GIS software, the latter part of the learning equation is undermined.

Past users have been in a similar situation of learning on a budget or staving off expensive software purchases until project requirements and full system benefits became apparent. The need for inexpensive or totally free GIS software surfaced in the early 1990s, and over the many years a catalog of quality, free GIS environments developed. In many cases, the free GIS software, called *freeware*, is a scaled-down version of the more popular fee-based software. These scaled-down versions are sometimes as basic as GIS viewers and ESRI-based shapefile editors, while other versions are fully functional base GIS systems. Both forms of freeware offer the novice GIS user a surefire means to gain the hands-on skills needed for proper understanding without a hefty monetary investment.

Answering this call for a free, fully functional GIS environment is GRASS, one of the oldest and most recognized GIS software applications available. GRASS, which stands for Geographic Resources Analysis Support System, was originally developed by the U.S. Army Construction Engineering Research Laboratories (USACERL), a branch of the U.S. Army Corps of Engineers, in 1982 as a military tool for land management and environmental planning. The USACERL maintained and revised the software up until 1995, at which time the software was deemed well developed and suitable for a wide range of applications. Today, GRASS is maintained by Baylor University in Waco, Texas, and has various mirror sites throughout North America and Europe.

By design, GRASS is a raster and topological vector-based GIS that supports image processing, geographic data management, spatial modeling, map and graphics production, visualization, and geospatial analysis. The software, with its straightforward functions and solid architecture, has gained wide appeal throughout the industry, whereby users have chosen GRASS over one of the more popular for-purchase GIS environments. The software contains more than 350 programs that offer add-on functions, such as raster/vector manipulation and enhanced mapping capabilities.

The GRASS GIS environment is used for academic and commercial applications worldwide by public and private organizations alike. In fact, the software is used by the National Aeronautics and Space Administration (NASA), National Oceanic and Atmospheric Administration (NOAA), U.S. Department of Agriculture (USDA), National Park Service (NPS), U.S. Census Bureau, and USGS. With wide industry use and public domain status, GRASS proves to be an excellent starting point for beginning GIS users.

As previously indicated, all types of skill sets have turned to GRASS and other free packages to produce viable GIS products. The demographics for system development and use are interesting. Free GIS software packages have been developed by university researchers, programmers, philanthropists, geospatial organizations, governmental agencies, and private developers since the 1980s. As with any mass assortment of developed software, a large number of the available titles were created to serve specific roles, either in application (e.g., mining and business statistics) or in function (e.g., file format conversions and image analysis). Beyond these specific titles are the base geospatial environments (such as GRASS) and standardized software that enable users on all platforms and of all skill sets to develop, manage, analyze, and map geographic data.

To guide you in this maze of freeware, Table 15.2 offers an abridged listing of popular free GIS environments, viewers, and shapefile editors available. More comprehensive listings of geospatial freeware exist on the Web through FreeGIS.org (http://www.freegis.org), GIS Lounge (http://www.gislounge.com), and the Open Geospatial Consortium (OpenGIS; http://www.opengis.org). You will be amazed at the amount of free GIS software available. In fact, be prepared when you visit FreeGIS.org; the site is astonishingly broad and has a cavernous array of good GIS software links.

The quality of freeware ranges from the very good to the moderately bad, which really means from high operability to difficult to use. Since GIS outputs

Table 15.2
Free GIS Environments, Viewers, and Shapefile Editors

ArcExplorer	http://www.esri.com/software/arcexplorer/index.html
AGISMap	http://www.agismap.com
Chameleon	http://chameleon.maptools.org
deegree	http://deegree.sourceforge.net
DIVA-GIS	http://www.diva-gis.org
Forestry GIS (fGIS)	http://www.digitalgrove.net/fgis.htm
GeoMedia Viewer	http://www.intergraph.com/gviewer
Geo/SQL	http://www.geosql.com
GRASS	http://www.baylor.edu/~grass
JGRASS	http://www.hydrologis.com/html/jgrass/jgrass_en.html
MapWindow GIS	http://www.mapwindow.com
idctlparmonoGIS	http://www.monogis.org
SAGA GIS	http://www.saga-gis.uni-goettingen.de/html/index.php
SPRING	http://www.dpi.inpe.br/spring/english/
TatukGIS Viewer	http://www.tatukgis.com/products/viewer/viewer.aspx
uDig	http://udig.refractions.net/confluence/display/UDIG/Home

are directly related to the geodata entered and not the system per se, quality really can be judged only on ingrained capability and functionality.

GIS Component Software

Once you begin using GIS software, whether free or purchased, you will soon realize that the software has limits. Most times the software will be satisfactory within these limits; other times you may want more. Perhaps the limit breaker is a data set that is in a different format that will take twice as long to integrate into your system or maybe a specific map analysis function that currently doesn't exist in the software. When more is desired, you have two options: Work around the limitations of your GIS environment or turn to GIS component software.

GIS component software is a building block that, when added to GIS software, forms an enhanced, personalized environment for the user. A mix of component software will make any GIS environment distinct for that user—making the environment the user's own. GIS component software allows the user to go far beyond the base system boundaries, limited only by a user's ability.

Component software is analogous to an office work desk. The actual furniture is like your GIS environment software—solid structure, yet standard for a multitude of users. The user's belongings—family pictures, reference books, or project paperwork—relative to the desk, just like component software to the GIS, make it the user's own. When used correctly, these belongings (components) add to the overall experience, use, and personalization of the desk (GIS environment) for a distinct user.

GIS component software comes in many variations, including specific function programs or user development software. Specific function components perform a dedicated task that adds to the GIS environment's tools. Such components include data format converters, flow data analyzers, and image processing software. User development software, on the other hand, is a development toolkit that enables the user to program components to perform specific functions. Say, for instance, you need to embed maps into a particular non-GIS program and there is no predeveloped component for your raster-only GIS. Developable component software would be the likely choice, especially if this operation is to be done numerous times. Simply put, component software helps save user time and automate redundant functions.

If component software interests you, there are many excellent products on the market. Perhaps the most popular component software available is Geotools (http://www.geotools.org), an open source GIS development toolkit that is freely distributed. The software is Java based and offers users the ability to develop OpenGIS-compliant products. OpenGIS is dedicated to developing and standardizing geospatial and geoprocessing specifications. What makes Geotools attractive to users is its modular design, which allows easy installation and removal of components. The software works well with the major fee-based and free GIS environments mentioned previously.

There are various component software and development toolkits available. Table 15.3 lists a variety of useful component software links for you to explore. Sites involve fee-based and free software. Care should be taken when selecting components to download, since not all components work with every GIS environment.

Once you get hands-on familiarity with a GIS environment and are comfortable with the base functions, you should try a component. Test out a few different components to decide what these building blocks can do for your efficiency and total GIS experience. In the end, only you can decide what (or if) components are right for you. Sometimes it depends just on the objective of a project. But as with your office desk, only you can decide what you need within arm's reach.

Table 15.3
GIS Component Software

ALOV Map/TMJava	http://www.alov.org/index.html
Blue Marble Geographics	http://www.bluemarblegeo.com
FlowMap	http://flowmap.geog.uu.nl
FME objects	http://www.safe.com/products/developer-tools/fme-objects/index.php
Generic Mapping Tools (GMT)	http://gmt.soest.hawaii.edu
GeoTools	http://www.geotools.org
JShape software	http://www.jshape.com/frame.jsp
MapE Library	http://www.mapesoft.com/MapELibrary.htm
Map Fusion	http://www.geosolutions.com/globalgeo/default.aspx
MapObjects	http://www.esri.com/getting_started/developers/mapobjects.html
MapX	http://extranet.mapinfo.com/products/Overview.cfm?productid=1041&productcategoryid=1
Open Map	http://openmap.bbn.com
SpatialFX	http://www.objectfx.com/products/spatialfx.asp
SylvanMaps/CF	http://www.sylvanmaps.com/sylvanmapsCF.htm
TatukGIS Developer Kernel	http://www.tatukgis.com/products/Dk/kernel.aspx

Geospatial Data Clearinghouses

It can be said that a GIS system is only as good as the user's ability and the data set used. No discussion on how to get started with GIS should dismiss the acquisition of geospatial data. There are fundamentally two ways to obtain geospatial data: creating it yourself or acquiring previously developed data sets. Creating a quality data set is neither simple nor cheap, as one would expect. Quality dataset development requires, at a minimum, field visits and/or field personnel, satellite image analysis, aerial image analysis, GPS coordinate attainment, geospatial data entry, preexisting dataset investigations, and any existing basemaps. The effort expended on creating custom data sets can be arduous and expensive; therefore, often firms or individuals creating the data set sell their product on the market or create a geodata "bi-product" to sell. You can refer to our discussion on data bi-products in Chapter 2.

Acquiring a previously developed data set is frequently more practical and, as such, the path most GIS users take. You can acquire much needed data instantly over the Internet either at a price that is far lower than creating the data set yourself or through freely distributed data sets. For specific data sets, perhaps the 2005 bushfires in Yellowstone National Park, previous data sets may not be available and the burden of creating it is on the user.

Previously created base and bi-product data sets are distributed primarily through geospatial data clearinghouses, data warehouses, and data depots. The terms *clearinghouse, warehouse,* and *depot* are nearly synonymous with each other. For our purposes, we will view them similarly, as collections and collaborative sources of geospatial data sets. These clearinghouses sell for-purchase data sets as well as distribute free data sets under a public domain license. Easier-to-obtain datasets, such as satellite-based data, are distributed free and less available data, such as detailed bi-products about specific areas, are often fee based.

Geospatial data clearinghouses are widely used throughout the numerous industries using GIS. As a learning GIS user, you should visit these clearinghouses quite frequently and obtain free data sets and metadata with which to hone your GIS skills. Advanced users frequent these clearinghouses regularly for their project maps and geographic information products, as well as for background metadata. Whatever the purpose, these data clearinghouses are among the best places to obtain viable geospatial data and metadata.

Table 15.4 details only a small handful of international data clearinghouses. All of these sites have free data sets, and some, such as the Data Depot, are totally dedicated to free data and metadata. Bear in mind that these are just a selection of the many hundreds of available clearinghouses. For a more comprehensive listing of data sites, refer to the University of Edinburgh-Association of Geographic Information's GIS Resource list (http://www.geo.ed.ac.uk/home/giswww.html) for links to several hundred U.S. and international GIS data sites.

Proper use of free (or any) geospatial data sets is essential for attaining effective and accurate GIS products. Users must carefully consider and understand the guidelines about how the data were obtained and how they should be used. This information lies within the data set's geospatial metadata, which we discussed back in Chapter 4. No matter whom the data comes from or what typical parameters are normally used, each data set is individual and should be treated as such for the sake of your product's quality and applicability. Assume nothing and keep the data quality bar raised.

At last, you are ready to turn theory into practice within the world of geospatial information and GIS. With the mystery and uncertainty about GIS removed by understanding, you are well on your way to creating geospatial answers to today's worldwide geographic questions. Armed with these discussions and the starting knowledge of the available GIS environments, you can keenly get yourself started with GIS.

Table 15.4
Geospatial Data Clearinghouses

The Australian Consortium for the Asian Spatial Information and Analysis Network	http://www.asian.gu.edu.au/index.html
Blue Marble Geographics World Map Dataset	http://www.bluemarblegeo.com/products/worldmapdata.php
The Data Store	http://www.data-store.co.uk
EROS Data Center	http://edc.usgs.gov
Federal Geographic Data Committee Geospatial Clearinghouse	http://www.fgdc.gov/clearinghouse/clearinghouse.html
GEO-DATA Explorer (GEODE)	http://geode.usgs.gov
The Geography Network	http://www.geographynetwork.com
Geo-Gratis	http://geogratis.cgdi.gc.ca/clf/en
Geoscience Australia National Geoscience Datasets	http://www.ga.gov.au/map/broker/wms_info.php
GIS Data Depot	http://data.geocomm.com
MapMart	http://www.mapmart.com
ResMap: Earth Image Source	http://www.resmap.com
United Nations Environment Network	http://www.unep.net
USDA Forest Service geospatial data clearinghouse	http://svinetfc4.fs.fed.us
USGS geographic data download	http://edc.usgs.gov/geodata
University of Edinburgh Data Library	http://datalib.ed.ac.uk/holdings.html

Afterword

The Future of the Growing GIS Technical Revolution

It is hard to believe that GISs have been around for more than four decades. Within this time frame, the world has experienced rather turbulent and exciting times, wherein society has dealt with human inequality issues, a superpower Cold War and escalating world unrest side by side (almost hand in hand) with the proliferation of the Internet, powerful microprocessors, and terrestrial and extraterrestrial exploration. Today, the thought of GIS as a "fresh" technology still exists, but in reality it has been reinvented and revised so many times that it now appears as something much different from its original form.

The first 40 years of GIS have proved that the core theory has remained unchanged, but the method and functionality of manipulating this theory and producing informative products have drastically changed. When GIS first came into being, it was developed and used primarily for land use issues. Its simple database structure offered a repository of land-based information and analysis. Through the years came the availability of satellite imagery, relational databases, desktop computers, powerful software, and world data infrastructures, just to name a few. Add these innovations with users who are better educated and have a multidisciplinary interest and the sum is an established, reinfused technology.

You may be asking yourself, "Where does GIS go from here?" or "How can this already established technology survive in the years to come?" Every user most likely has a personal opinion as to the future of GIS. In fact, there has been a good amount of print and online space dedicated to the dissemination of opinion and prediction.

A common theme throughout many opinion papers is the belief that history is the best indicator for technological evolution, since it reconstitutes in continuous yet variant forms. I hold this same belief. Consequently, if history proves a valid gauge for GIS, then the technology will ultimately survive through further revision and enhanced redirection. Although this sounds as generic a prediction as a newspaper horoscope, the mere fact that GIS still exists today and is a developed technology adds value to this claim.

Let us move beyond today's version of GIS and postulate about the direction it is going. For this, I have developed what I call a GIS compass. This GIS compass (and this entire discussion for that matter) represents my own personal musings as to the direction of GIS. The following compass directions are selections of real issues that, to me, may influence the course of GIS's evolution.

The Developing GIS Compass

A compass is a worthy instrument, used to navigate terrains and help the user get from a point of origin to one of destination. It also helps the user discern what direction he or she is going. Understanding the value of a compass and the need to gauge the GIS industry's direction, I created the GIS compass. My objective was to help convey a selection of emerging terrains for GIS and identify the chief directions that I believe the industry is currently moving. Figure A.1 depicts the developing GIS compass.

As evident through today's GIS use, the industry is quickly moving in four distinct and primary directions. I define them as:

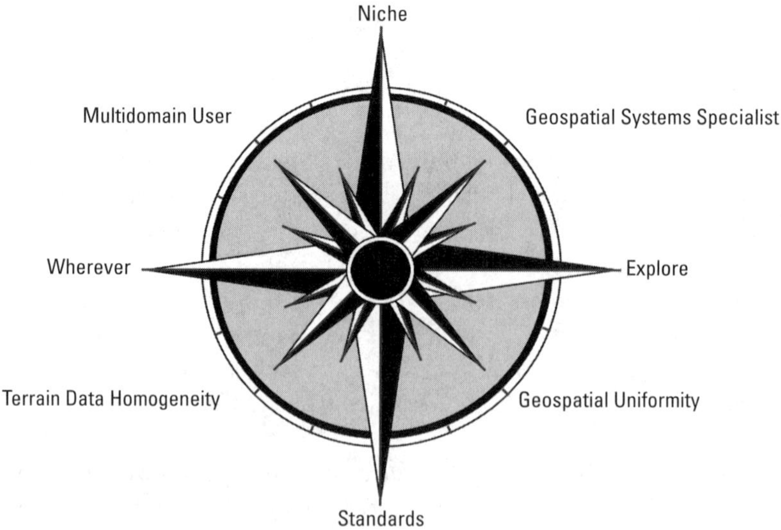

Figure A.1 The GIS compass.

- *N for niche.* Moving from the present-day GIS professional to an information technology niche field;
- *E for explore.* Moving from a geodatabase to enhanced geospatiality by exploring the increasing geospatial landscape;
- *W for wherever.* Progressing from the Earth to other topographies (both macroscopic and microscopic), such as interstellar terrains and inner body mapping;
- *S for standards.* Developing universally respected global standards.

These primary directions are speculative in nature and were determined through my interactions in the industry. User types are diversifying, applications are broadening, and the knowledge that any mapable surface can be geospatially coded is spreading. Among these primary directions are meldings in between that, I believe, will ultimately form the specifications to industry survival. Let us now take a brief look at these key pathways.

Niche

The GIS professional was always just that, a professional dedicated to GIS. These dedicated persons form the core staff responsible for (and knowledgeable about) spatially enabled geodatabases. Typically, they are the project members who provide the geospatial grunt work, such as uploading basemaps and data, validating data, geocoding resources, troubleshooting geospatial inaccuracies, and producing specific information products. They are the dedicated professionals for GIS.

With greater interest and better software abounding, it appears that today's GIS professionals are sometimes sidetracked by GIS do-it-yourselfers or project managers who skimp on their budgets. It is not out of the ordinary for GIS professionals to be tasked with other non-GIS project responsibilities. Most commonly, the GIS professional is slated as the information technology (IT) specialist, wherein the workload consists of both geographically based and general information–based system tasks. It has been my experience that the GIS professional's role also involves the actual IT hardware configurations.

This leads us to our first GIS direction: niche. I believe the future GIS professionals will not be GIS professionals per se but IT specialists with a GIS or geospatial niche. The need for a dedicated geospatial person, at least in the engineering and technological fields, is already being thwarted by minimized budgets and a push for multitasking. This is not the case for every company. Many large companies have teams of GIS professionals providing services that have sustained over time, and for federal and local agencies, the need for dedicated GIS professionals will undoubtedly remain.

Like everything else, only time will determine to what extent and in what sectors the industry will go in this direction. One thing is clear, though: that the upcoming GIS professionals will need to diversify their capabilities in order to remain valid in the marketplace.

Explore

As GIS moves toward and into every feasible technical and nontechnical crevice, the understanding (even enlightenment) of geospatiality continues to occur. When GIS began, it was a geographic database; today it is a geospatial data infrastructure. The idea that environments can be geospatially linked to other information is exciting to most, especially when it offers forms of analysis that never existed before. We have school systems exploring the geospatial terrain in their districts, such as crime statistics, neighborhood demographics, district and nondistrict school information, and more. Today's industries are just beginning to understand the absolute value of geospatial data and how they can use that information to make better decisions and grow their businesses.

Agencies have traditionally been the geospatial leaders and are continuing to broaden their range of analysis. GIS has progressed from land use applications to such exciting applications as geospatially referenced rainforest inventories. Homeland security has emerged in the geospatial realm, using available and project-acquired data to ensure public safety, while remaining a few steps ahead of the malcontents.

Exploring the geospatial landscapes involves increased knowledge of how to use and manipulate data. Geospatial training is already a large market in today's marketplace and, I trust, will likely grow into a huge on-demand market. More firms and agencies will likely continue to invest money for geospatial awareness training, since the potential cost savings and advantages to proprietary information are great.

Wherever

Similar to GIS's "explore" direction is a trend that is voraciously gaining speed: moving from Earth's immediate surface to diverse terrains and topographies. GIS has traditionally been applied to Earthbound geography, but in past years GIS has emerged as the system of choice for various different topographies. For instance, NASA has used GIS to precisely map and conduct unique analysis of the Moon's surface characteristics and, with the aid of its two robotic Martian rovers, has geospatially mapped the surface of Mars.

This "wherever" direction is not necessarily macroscopic or solely applicable to extraterrestrial landscapes; it also applies to landscapes too small to see. The medical industry is in its infancy in using GIS technology. I'm not talking about disease tracking and prevention analysis, but rather the inner landscapes of the body, such as geocoding a map of the human brain or the human body's constituents.

The scientific community is progressing with GIS, as well. The potential for a union of underwater microscopy and GIS has only begun to be explored. I foresee GIS being used to monitor and analyze the various life forms and living structures on Earth's declining coral reefs. The absolute potential for once unimaginable analysis is made available through geospatial technology.

But more must be done to move farther in this primary direction. As stated previously, training will play a giant role in progressing industry understanding. The medical and scientific fields are just two of the myriad disciplines GIS can benefit. The tricky part, however, is getting GIS integrated within these sectors. If that weren't complicated enough, professionals must work "outside the intellectual box" and be creative with their applications in order to expand the overall system benefits. Now, I agree that GIS may not prove to be the be all and end all tool, but I do think it will play a growing role in many fields new to GIS. As with any technology, the overall efficiency and usefulness of the technology is intimately dependent upon the user's capability.

Standards

An area where there is presently the greatest amount of activity and conflict is with GIS standards. Today's GIS industry presents numerous and varying standards that are supported by some organizations and disputed by others. The procedures, protocol, format, and implementation of various GIS elements, such as metadata, are continuously experiencing change and redirection. In fact, there are some geospatial standards prominent in the United States that differ from the prominent standards in Europe and Australia.

As a promising pathway for geospatial consistency, some globally focused organizations are gaining the support and validation needed to create new standards and fluid interoperability. In recent years, the Open Geospatial Consortium has made great strides in creating the OpenGIS, a global system dedicated to universal GIS compatibility. I recognize that defining global standards will afford all GISs and geospatial data sets a method to seamlessly interoperate with each another.

Similarly, there are numerous metadata formats that are progressing toward global standardization, such as the FGDC, Dublin Core Initiative, and ISO efforts. The Open Geospatial Consortium has even posted guidelines for metadata, which has proved to be one stepping stone for international metadata standardization. With each metadata schema, there are very different minimum requirements, in addition to differing metadata semantics and syntax. I foresee this direction progressing further.

Even though great strides have been made through the ISO metadata efforts, I sense that global standards may never occur, simply because it would require a good many organizations worldwide to buy into the structure, discipline, and protocol of a specific standard. If these universal standards are achievable, it will most likely be countless years before global standards truly manifest and get approval by the whole GIS community. That being stated, if this direction's destination is reached, it has the potential of redefining the way GIS is done in the years to come.

All Directions in Between

The GIS compass is unique in that between my primary anticipated directions (refer to Figure A.1) there are meldings—geospatial systems specialist, geospatial uniformity, terrain data homogeneity, and multidomain user—that offer likely products. The geospatial systems specialist, for instance, is a melding of two directions—a niche IT person specializing in all things geospatial. Likewise, geospatial uniformity, to me, involves a melding of geospatial exploration (all things geospatial) and standardization.

Terrain data homogeneity melds the idea of any terrain (on or off the Earth) with standards and homogeneous use of the data. The multidomain user direction melds the notion of using GIS wherever necessary with specialized (niche) yet multidisciplinary users.

As you may have realized already, it is my opinion that the progression of the four primary directions can be met through creative combinations of each. These combinations are manifesting today worldwide and, as a guess, will increase in both occurrence and availability. It will not be too long before doctors and building managers alike are using GIS to conduct business tasks, or geospatial products from different continents are cataloged the same exact way throughout the world. Apparently, there will also be a time in the not too distant future when GIS systems will read all formats without fail. Right now, these aspirations are in process and remain to fully manifest. I feel, given time, they will. You can say they are whims of fancy and hold no certainty as yet; however, one thing is definite: that GIS remains a true technical revolution.

Appendix A: Reference Ellipsoid Parameters

Considered a worldwide ellipsoid, WGS84 is one of the most widely used reference ellipsoids available. However, there are several major ellipsoids that are focused to specific areas and more locally used. Table A.1 highlights key reference ellipsoids and provides the associated semimajor axis (equatorial radius), semiminor axis (polar radius), and inverse flattening parameters. Please note that semiminor axis constants were calculated using the following formula:

$$\textit{Semiminor axis} = \textit{Semimajor axis} \times \left(1 - \frac{1}{\textit{Inverse flattening}}\right)$$

Table A.1
Reference Ellipsoid Parameters

Ellipsoid Name	Semimajor Axis (meters)	Semiminor Axis (meters)	Inverse Flattening (f^{-1})
Airy 1830	6,377,563.396	6,356,256.909	299.3249646
APL 4.5 (1968)	6,378,144	6,356,757.339	298.23
Australian National	6,378,160	6,356,774.719	298.25
Average Terrestrial System 1977	6,378,135	6,356,750.305	298.257
Bessel 1841 (Ethiopia, Indonesia, Japan, and Korea)	6,377,397.155	6,356,078.963	299.1528128
Bessel 1841 (Namibia)	6,377,483.865	6,356,165.383	299.1528128

Table A.1 (continued)

Clarke 1858	6,378,235.6	6,356,560.14	294.2606768
Clarke 1858–Modified	6,378,293.645	6,356,617.938	294.26
Clarke 1866	6,378,206.4	6,356,583.8	294.9786982
Clarke 1880	6,378,249.145	6,356,514.87	293.465
Danish–Andrae (1876)	6,377,104.43	6,355,847.415	300
Delambre 1810	6,376,985.228	6,356,323.664	308.64
Everest	6,377,276.345	6,356,075.413	300.8017
Everest (India 1830)	6,377,276.345	6,356,075.413	300.8017
Everest (India 1956)	6,377,301.243	6,356,100.228	300.8017
Everest (Pakistan)	6,377,309.613	6,356,108.571	300.8017
Everest (Sabah and Sarawak)	6,377,298.556	6,356,097.55	300.8017
Everest (W. Malaysia 1969)	6,377,295.664	6,356,094.668	300.8017
Everest (W. Malaysia and Singapore 1948)	6,377,304.063	6,356,103.039	300.8017
Fischer 1960 (Mercury)	6,378,166	6,356,784.284	298.3
Fischer 1968	6,378,150	6,356,768.337	298.3
GEM 10C	6,378,137	6,356,752.314	298.2572236
Germaine (Djibouti)	6,378,284	6,356,589.156	294
GRS67	6,378,160	6,356,774.516	298.2471674
GRS75	6,378,140	6,356,755.288	298.257
GRS80	6,378,137	6,356,752.314	298.2572236
Helmert 1906	6,378,200	6,356,818.17	298.3
Hough 1960	6,378,270	6,356,794.343	297
Indonesian 1974	6,378,160	6,356,774.504	298.247
International 1924	6,378,388	6,356,911.946	297
Krassovsky 1940	6,378,245	6,356,863.019	298.3
Krayenhoff 1827	6,376,950.4	6,356,356.341	309.65
Modified Airy 1849	6,377,340.189	6,356,034.448	299.3249646
Modified Everest	6,377,304.063	6,356,103.039	300.8017
Modified Fischer 1960	6,378,155	6,356,773.32	298.3
NWL 10D	6,378,135	6,356,750.52	298.26
NWL 9D	6,378,145	6,356,759.769	298.25
OSU 86F	6,378,136.2	6,356,751.517	298.25722
OSU 91A	6,378,136.3	6,356,751.616	298.25722
Plessis Modified 1817	6,376,523	6,355,862.933	308.64
Plessis Reconstituted (Modified DeLambre 1810)	6,376,523.994	6,355,862.907	308.624807

Table A.1 (continued)

SGS 85	6,378,136	6,356,751.302	298.257
SGS 90	6,378,136	6,356,751.362	298.2578393
South American 1969	6,378,160	6,356,774.719	298.25
Struve 1860	6,378,298.3	6,356,657.143	294.73
Svanberg	6,376,797	6,355,837.971	304.2506
Walbeck 1819–AMS (1963)	6,376,896	6,355,834.847	302.78
Walbeck 1819–Planheft (1942)	6,376,895	6,355,834	302.7821565
War Office 1926 (McCaw 1924)	6,378,300	6,356,751.689	296
WGS60	6,378,165	6,356,783.287	298.3
WGS66	6,378,145	6,356,759.769	298.25
WGS72	6,378,135	6,356,750.52	298.26
WGS84	6,378,137	6,356,752.314	298.2572236
Xian 1980	6,378,140	6,356,755.288	298.257

Appendix B: Worldwide Geodetic Datums

Using geodetic datums correctly is an important part of working with GIS and producing accurate information products. Although similar, geodetic datums are distinct, having unique geodetic characteristics. Therefore, conversion between datums is needed. To convert from one datum to another, specific conversion parameters and the three-dimensional coordinate shifts between both datums must be known. In more complex conversions, where coordinate axes are not parallel and/or scaled, coordinate axes rotational parameters and a scale difference factor must be known as well.

Since the World Geodetic System 1984 uses an Earth-centered, Earth-fixed coordinate system, it is a highly used, worldwide datum (and reference ellipsoid). In fact, GPS is based upon WGS84. Due to its worldwide appeal, many applications require local datums to be converted to WGS84. As with any simple geodetic datum transformation, basic datum transformation parameters must be known.

Table B.1 contains a brief listing of geodetic datum transformation parameters relative to WGS84. It highlights some of the more conventional datums used throughout the world, but by no means is it a comprehensive listing of available datums. Further information and more complete local listings are freely distributed through the datum sources cited.

The following base parameters were used to create Table B.1:

$$WGS84\ semimajor\ axis = 6{,}378{,}137$$

$$WGS84\ flattening = 0.003352811$$

Table B.1
Worldwide Geodetic Datums

Location	Datum Name	Ellipsoid	Semimajor Axis (meters)	Flattening (f)	Datum Shift X	Datum Shift Y	Datum Shift Z
Africa							
Algeria	North Sahara 1959	Clarke 1880	6,378,249.145	0.003407561	–186	–93	310
Algeria	Voirol 1960	Clarke 1880	6,378,249.145	0.003407561	–123	–206	219
Botswana	ARC 1950	Clarke 1880	6,378,249.145	0.003407561	–138	–105	–289
Burkina Faso	Adindan	Clarke 1880	6,378,249.145	0.003407561	–118	–14	218
Burkina Faso and Niger (mean)	Point 58	Clarke 1880	6,378,249.145	0.003407561	–106	–129	165
Burundi	ARC 1950	Clarke 1880	6,378,249.145	0.003407561	–153	–5	–292
Cameroon	Adindan	Clarke 1880	6,378,249.145	0.003407561	–134	–2	210
Cameroon	Minna	Clarke 1880	6,378,249.145	0.003407561	–81	–84	115
Congo	Pointe Noire 1948	Clarke 1880	6,378,249.145	0.003407561	–148	51	–291
Djibouti	Ayabelle Lighthouse	Clarke 1880	6,378,249.145	0.003407561	–79	–129	145
Egypt	Old Egyptian 1907	Helmert 1906	6,378,200	0.00335233	–130	110	–13
Egypt	European 1950	International 1924	6,378,388	0.003367003	–130	–117	–151
Ethiopia	Adindan	Clarke 1880	6,378,249.145	0.003407561	–165	–11	206
Ethiopia (Eritrea)	Massawa	Bessel 1841 (Ethiopia, Indonesia, Japan, and Korea)	6,377,397.155	0.003342773	639	405	60
Gabon	M'Poraloko	Clarke 1880	6,378,249.145	0.003407561	–74	–130	42
Ghana	Legion	Clarke 1880	6,378,249.145	0.003407561	–130	29	364
Guinea	Dabola	Clarke 1880	6,378,249.145	0.003407561	–83	37	124
Guinea–Bissau	Bissau	International 1924	6,378,388	0.003367003	–173	253	27

Table B.1 (Continued)

Kenya	ARC 1960	Clarke 1880	6,378,249.145	0.003407561	−157	−2	−299
Lesotho	ARC 1950	Clarke 1880	6,378,249.145	0.003407561	−125	−108	−295
Liberia	Liberia 1964	Clarke 1880	6,378,249.145	0.003407561	−90	40	88
Madagascar	Tananarive Observatory 1925	International 1924	6,378,388	0.003367003	−189	−242	−91
Malawi	ARC 1950	Clarke 1880	6,378,249.145	0.003407561	−161	−73	−317
Mali	Adindan	Clarke 1880	6,378,249.145	0.003407561	−123	−20	220
Morocco	Merchich	Clarke 1880	6,378,249.145	0.003407561	31	146	47
Namibia	Schwarzeck	Bessel 1841 (Namibia)	6,377,483.865	0.003342773	616	97	−251
Nigeria	Minna	Clarke 1880	6,378,249.145	0.003407561	−92	−93	122
Senegal	Adindan	Clarke 1880	6,378,249.145	0.003407561	−128	−18	224
Sierra Leone	Sierra Leone 1960	Clarke 1880	6,378,249.145	0.003407561	−88	4	101
Somalia	Afgooye	Krassovsky 1940	6,378,245	0.00335233	−43	−163	45
South Africa	Cape	Clarke 1880	6,378,249.145	0.003407561	−136	−108	−292
Sudan	Adindan	Clarke 1880	6,378,249.145	0.003407561	−161	−14	205
Swaziland	ARC 1950	Clarke 1880	6,378,249.145	0.003407561	−134	−105	−295
Tanzania	ARC 1950	Clarke 1880	6,378,249.145	0.003407561	−175	−23	−303
Tunisia	Carthage	Clarke 1880	6,378,249.145	0.003407561	−263	6	431
Tunisia	European 1950	International 1924	6,378,388	0.003367003	−112	−77	−145
Zaire	ARC 1950	Clarke 1880	6,378,249.145	0.003407561	−169	−19	−278
Zambia	ARC 1950	Clarke 1880	6,378,249.145	0.003407561	−147	−74	−283
Zimbabwe	ARC 1950	Clarke 1880	6,378,249.145	0.003407561	−142	−96	−293
Antarctica							
Camp McMurdo Area	Camp Area Astro	International 1924	6,378,388	0.003367003	−104	−129	239
Deception Island	Deception Island	Clarke 1880	6,378,249.145	0.003407561	260	12	147

Table B.1 (Continued)

Asia							
Afghanistan	Herat North	International 1924	6,378,388	0.003367003	−133	−222	114
Bahrain Island	Ain el Abd 1970	International 1924	6,378,388	0.003367003	−150	−250	−1
Bangladesh	Indian	Everest (India 1830)	6,377,276.345	0.003324449	282	726	254
Brunei and East Malaysia	Timbalai 1948	Everest (Sabah and Sarawak)	6,377,298.556	0.003324449	−679	669	−48
Con Son Island (Vietnam)	Indian 1960	Everest (India 1830)	6,377,276.345	0.003324449	182	915	344
Hong Kong	Hong Kong 1963	International 1924	6,378,388	0.003367003	−156	−271	−189
India and Nepal	Indian	Everest (India 1956)	6,377,301.243	0.003324449	295	736	257
Indonesia	Indonesian 1974	Indonesian 1974	6,378,160	0.003352926	−24	−15	5
Indonesia (Sumatra)	Djakarta	Bessel 1841 (Ethiopia, Indonesia, Japan, and Korea)	6,377,397.155	0.003342773	−377	681	−50
Iran	European 1950	International 1924	6,378,388	0.003367003	−117	−132	−164
Iwo Jima	Astro Beacon "E" 1945	International 1924	6,378,388	0.003367003	145	75	−272
Japan, South Korea, and Okinawa (mean)	Tokyo	Bessel 1841 (Ethiopia, Indonesia, Japan, and Korea)	6,377,397.155	0.003342773	−148	507	685
Okinawa	Tokyo	Bessel 1841 (Ethiopia, Indonesia, Japan, and Korea)	6,377,397.155	0.003342773	−158	507	676
Oman	Oman	Clarke 1880	6,378,249.145	0.003407561	−346	−1	224
Philippines (except Mindanao Island)	Luzon	Clarke 1866	6,378,206.4	0.003390075	−133	−77	−51
Philippines (Mindanao Island)	Luzon	Clarke 1866	6,378,206.4	0.003390075	−133	−79	−72
Qatar	Qatar National	International 1924	6,378,388	0.003367003	−128	−283	22
Republic of Maldives	Gan 1970	International 1924	6,378,388	0.003367003	−133	−321	50
Russia	S42 (Pulkova 1942)	Krassovsky 1940	6,378,245	0.00335233	28	−130	−95
Saudi Arabia	Nahrwan	Clarke 1880	6,378,249.145	0.003407561	−243	−192	477
Saudi Arabia	Ain el Abd 1970	International 1924	6,378,388	0.003367003	−143	−236	7
Singapore	South Asia	Modified Fischer 1960	6,378,155	0.00335233	7	−10	−26

Table B.1 (Continued)

South Korea	Tokyo	Bessel 1841 (Ethiopia, Indonesia, Japan, and Korea)	6,377,397.155	0.003342773	–147	506	687
Sri Lanka	Kandawala	Everest (India 1830)	6,377,276.345	0.003324449	–97	787	86
Taiwan	Hu-Tzu-Shan	International 1924	6,378,388	0.003367003	–637	–549	–203
Thailand	Indian 1954	Everest (India 1830)	6,377,276.345	0.003324449	217	823	299
United Arab Emirates	Nahrwan	Clarke 1880	6,378,249.145	0.003407561	–249	–156	381
Vietnam (near 16° north)	Indian 1960	Everest (India 1830)	6,377,276.345	0.003324449	198	881	317
West Malaysia and Singapore	Kertau 1948	Everest (W. Malaysia and Singapore 1948)	6,377,304.063	0.003324449	–11	851	5
Atlantic Ocean							
Greenland	NAD27	Clarke 1866	6,378,206.4	0.003390075	11	114	195
Greenland (South)	Qornoq	International 1924	6,378,388	0.003367003	164	138	–189
Iceland	Hjorsey 1955	International 1924	6,378,388	0.003367003	–73	46	–86
Puerto Rico and Virgin Islands	Puerto Rico	Clarke 1866	6,378,206.4	0.003390075	11	72	–101
Trinidad and Tobago	Naparima, BWI	International 1924	6,378,388	0.003367003	–10	375	165
Australia							
Australia and Tasmania	Australian Geodetic 1966	Australian National	6,378,160	0.003352892	–133	–48	148
Australia and Tasmania	Australian Geodetic 1984	Australian National	6 378,160	0.003352892	–134	–48	149
Canada							
Alberta and British Columbia	NAD27	Clarke 1866	6.378,206.4	0.003390075	–7	162	188
Canada	NAD83	GRS80	6,378,137	0.003352811	0	0	0
Manitoba and Ontario	NAD27	Clarke 1866	6,378,206.4	0.003390075	–9	157	184
New Brunswick, Newfoundland, Nova Scotia, and Quebec	NAD27	Clarke 1866	6,378,206.4	0.003390075	–22	160	190

Table B.1 (Continued)

Northwest Territories and Saskatchewan	NAD27	Clarke 1866	6,378,206.4	0.003390075	4	159	188
Yukon	NAD27	Clarke 1866	6,378,206.4	0.003390075	–7	139	181
Caribbean and Central America							
Antigua, Barbados, Barbuda, Caicos Islands, Cuba, Dominican Republic, Grand Cayman, Jamaica and Turks Islands	NAD27	Clarke 1866	6,378,206.4	0.003390075	–3	142	183
Belize, Costa Rica, El Salvador, Guatemala, Honduras, and Nicaragua (mean)	NAD27	Clarke 1866	6,378,206.4	0.003390075	0	125	194
Canal Zone	NAD27	Clarke 1866	6,378,206.4	0.003390075	0	125	201
Cuba	NAD27	Clarke 1866	6,378,206.4	0.003390075	–9	152	178
Mexico	NAD27	Clarke 1866	6,378,206.4	0.003390075	–12	130	190
Mexico and Central America	NAD83	GRS80	6,378,137	0.003352811	0	0	0
Europe							
Albania	S42 (Pulkova 1942)	Krassovsky 1940	6,378,245	0.00335233	24	–130	–92
Austria, Denmark, France, West Germany, Netherlands, and Switzerland (mean)	European 1950	International 1924	6,378,388	0.003367003	–87	–96	–120
Austria, Finland, Netherlands, Norway, Spain, Sweden, and Switzerland (mean)	European 1979	International 1924	6,378,388	0.003367003	–86	–98	–119
Cyprus	European 1950	International 1924	6,378,388	0.003367003	–104	–101	–140
Czechoslavakia	S42 (Pulkova 1942)	Krassovsky 1940	6,378,245	0.00335233	26	–121	–78
rin0Czechoslavakia (prior to January 1, 1993)	S-JTSK	Bessel 1841 (Ethiopia, Indonesia, Japan and Korea)	6,377,397.155	0.003342773	589	76	480
England	Ordinance Survey of Great Britian 1936	Airy 1830	6,377,563.396	0.003340851	371	–112	434
England, Ireland, Scotland, Shetland Islands, and Channel Islands	European 1950	International 1924	6,378,388	0.003367003	–86	–96	–120

Table B.1 (Continued)

England, Wales, and Island of Man	Ordinance Survey of Great Britian 1936	Airy 1830	6,377,563.396	0.003340851	371	–111	434
Estonia	Coordinate System 1937 of Estonia	Bessel 1841 (Namibia)	6,377,483.865	0.003342773	374	150	588
Finland and Norway	European 1950	International 1924	6,378,388	0.003367003	–87	–95	–120
Greece	European 1950	International 1924	6,378,388	0.003367003	–84	–95	–130
Hungary	S42 (Pulkova 1942)	Krassovsky 1940	6,378,245	0.00335233	28	–121	–77
Ireland	Ireland 1965	Modified Airy 1849	6,377,340.189	0.003340851	506	–122	611
Italy	Rome 1940	International 1924	6,378,388	0.003367003	–225	–65	9
Italy (Sardinia)	European 1950	International 1924	6,378,388	0.003367003	–97	–103	–120
Italy (Sicily)	European 1950	International 1924	6,378,388	0.003367003	–97	–88	–135
Kazakhstan	S42 (Pulkova 1942)	Krassovsky 1940	6,378,245	0.00335233	15	–130	–84
Latvia	S42 (Pulkova 1942)	Krassovsky 1940	6,378,245	0.00335233	24	–124	–82
Malta	European 1950	International 1924	6,378,388	0.003367003	–107	–88	–149
Poland	S42 (Pulkova 1942)	Krassovsky 1940	6,378,245	0.00335233	23	–124	–82
Portugal and Spain	European 1950	International 1924	6,378,388	0.003367003	–84	–107	–120
Romania	S42 (Pulkova 1942)	Krassovsky 1940	6,378,245	0.00335233	28	–121	–77
Scotland and Shetland Islands	Ordinance Survey of Great Britian 1936	Airy 1830	6,377,563.396	0.003340851	384	–111	425
Wales	Ordinance Survey of Great Britian 1936	Airy 1830	6,377,563.396	0.003340851	370	–108	434
Yugoslavia (before 1990), Slovenia, Crotia, Bosnia, Herzegovina, and Serbia	Hermannskogel	Bessel 1841 (Namibia)	6,377,483.865	0.003342773	653	–212	449
Pacific Ocean							
American Samoa Islands	American Samoa 1962	Clarke 1866	6,378,206.4	0.003390075	–115	118	426
Easter Island	Easter Island 1967	International 1924	6,378,388	0.003367003	211	147	111
Guam	Guam 1963	Clarke 1866	6,378,206.4	0.003390075	–100	–248	259
Johnston Island	Johnston Island 1961	International 1924	6,378,388	0.003367003	189	–79	5
Marshall Islands	Wake-Eniwetok 1960	Hough 1960	6,378,270	0.003367003	102	52	–38
New Zealand	Geodetic Datum 1949	International 1924	6,378,388	0.003367003	84	–22	209

Table B.1 (Continued)

Viti Levu Island (Fiji Islands)	Viti Levu 1916	Clarke 1880	6,378,249.145	0.003407561	51	391	–36
South America							
Argentina	Campo Inchauspe	International 1924	6,378,388	0.003367003	–148	136	90
Argentina	South American 1969	South American 1969	6,378,160	0.003352892	–62	–1	–37
Baltra and Galapagos Islands	South American 1969	South American 1969	6,378,160	0.003352892	–47	26	–42
Bolivia	Provisional South American 1956	International 1924	6,378,388	0.003367003	–270	188	–388
Bolivia	South American 1969	South American 1969	6,378,160	0.003352892	–61	2	–48
Brazil	Corrego Alegre	International 1924	6,378,388	0.003367003	–206	172	–6
Brazil	South American 1969	South American 1969	6,378,160	0.003352892	–60	–2	–41
Chile	South American 1969	South American 1969	6,378,160	0.003352892	–75	–1	–44
Chile (Northern)	Provisional South American 1956	International 1924	6,378,388	0.003367003	–270	183	–390
Chile (Southern)	Provisional South American 1956	International 1924	6,378,388	0.003367003	–305	243	–442
Colombia	Bogota Observatory	International 1924	6,378,388	0.003367003	307	304	–318
Colombia	Provisional South American 1956	International 1924	6,378,388	0.003367003	–282	169	–371
Colombia	South American 1969	South American 1969	6,378,160	0.003352892	–44	6	–36
Ecuador	Provisional South American 1956	International 1924	6,378,388	0.003367003	–278	171	–367
Ecuador (excluding Galapagos Islands)	South American 1969	South American 1969	6,378,160	0.003352892	–48	3	–44
Guyana	Provisional South American 1956	International 1924	6,378,388	0.003367003	–298	159	–369
intblGuyana	South American 1969	South American 1969	6,378,160	0.003352892	–53	3	–47
Paraguay	Chua Astro	International 1924	6,378,388	0.003367003	–134	229	–29
Paraguay	South American 1969	South American 1969	6,378,160	0.003352892	–61	2	–33
Peru	Provisional South American 1956	International 1924	6,378,388	0.003367003	–279	175	–379
Peru	South American 1969	South American 1969	6,378,160	0.003352892	–58	0	–44
Suriname	Zanderij	International 1924	6,378,388	0.003367003	–265	120	–358
Uruguay	Yacare	International 1924	6,378,388	0.003367003	–155	171	37
Venezuela	Provisional South American 1956	International 1924	6,378,388	0.003367003	–295	173	–371

Table B.1 (Continued)

Venezuela	South American 1969	South American 1969	6,378,160	0.003352892	–45	8	–33
United States							
Alaska (except Aleutian Islands)	NAD27	Clarke 1866	6,378,206.4	0.003390075	–5	135	172
Alaska (except Aleutian Islands)	NAD83	GRS 80	6,378,137	0.003352811	0	0	0
Aleutian Islands	NAD83	GRS 80	6,378,137	0.003352811	–2	0	4
Bahamas (except San Salvador Island)	NAD27	Clarke 1866	6,378,206.4	0.003390075	–4	154	178
Continental U.S. (east of Mississippi River, including Louisiana, Minnesota, and Missouri)	NAD27	Clarke 1866	6,378,206.4	0.003390075	–9	161	179
Continental U.S. (west of Mississippi River, excluding Louisiana, Minnesota, and Missouri)	NAD27	Clarke 1866	6,378,206.4	0.003390075	–8	159	175
Florida and Bahamas (mean)	Cape Canaveral	Clarke 1866	6,378,206.4	0.003390075	–2	151	181
Hawaii	Old Hawaiian	Clarke 1866	6,378,206.4	0.003390075	89	–279	–183
Hawaii	NAD83	GRS 80	6,378,137	0.003352811	1	1	–1
Kauai, Hawaii	Old Hawaiian	Clarke 1866	6,378,206.4	0.003390075	45	–290	–172
Kauai, Hawaii	Old Hawaiian	International 1924	6,378,388	0.003367003	185	–233	–337
Maui, Hawaii	Old Hawaiian	Clarke 1866	6,378,206.4	0.003390075	65	–290	–190
Maui, Hawaii	Old Hawaiian	International 1924	6,378,388	0.003367003	205	–233	–355
Oahu, Hawaii	Old Hawaiian	Clarke 1866	6,378,206.4	0.003390075	58	–283	–182
Oahu, Hawaii	Old Hawaiian	International 1924	6,378,388	0.003367003	198	–226	–347

Glossary

absolute accuracy Map accuracy based on geographic coordinates and specific to an object's geographic location in relation to its true location on the surface of the Earth.

active pages Web pages that offer full user interaction, whereby the user can request information, as well as make changes to and manage the database.

aspect The relative orientation of the projected plane and the ellipsoid of revolution with respect to the location of an observer.

attributes Columns in a relational database management system (RDBMS) that contain distinct field data.

axis A line of symmetry within a shape.

axis of revolution *See* axis of rotation.

axis of rotation An axis upon which a two-dimensional shape is rotated to form a three-dimensional shape. For instance, an ellipse is rotated upon its major axis to form an ellipsoid.

azimuth The horizontal component of a line's direction, measured in degrees clockwise from the baseline; the baseline is typically the vertical axis (*y*-axis), and both the line and the baseline share a common center point.

azimuthal projection *See* planar projection

bar scale *See* graphic scale

bi-products Developable information products formed as a result of a specific application data process.

Cartesian coordinate system A two-dimensional, planar coordinate system, where horizontal and vertical distances are measured as x and y coordinates.

cartography The organization, design, collection, portrayal, and reproduction of geographically based information in various formats, such as on graphic maps or within digital formats.

Cassini, Jean-Dominique (1625–1712) An eighteenth-century patriarch of a renowned French geodesic family who erroneously concluded that the Earth pointed toward the poles and flattened at the equator. Cassini's Earth model was a prolate ellipsoid and resembled an egg shape. It was the favored model among the French geodetic community until the French Academy of Sciences proved him incorrect in 1745. Cassini was born Giovanni Domenico Cassini in Genoa, Italy, and after moving to France changed his name to Jean-Dominique.

centroid The center of an area, region, or polygon represented as discrete x-y coordinates.

circle A symmetrical two-dimensional shape in which any point on the outer shape is a set distance (radius) from a center point.

client-server system A relational database type that is designed to effectively handle with speed and efficiency the many users accessing the central database at one time. This system uses a network to connect the user workstations (called clients) to the central database residing on a file server.

Codd, Edgar F. (1923–2003) The author of "A Relational Model of Data for Large Shared Data Banks," who detailed a new method of simplifying database management using relational algebra for straightforward database functions. He is considered the forefather of the relational database management system. *See* Relational database management system (RDBMS)

compensatory scale factors Variant scale factors that compensate for one another to preserve a projection's area characteristic. These scale factors are typically greater than 1.0 in one direction and less than 1.0 in another direction. In equal-area projections, using compensatory scale factors will project a circle on a globe as an ellipse on a plane surface.

computer-aided design (CAD) Computer-based design software used to produce various technical and design drawings, specifically in the engineering, architectural, and environmental fields.

conformal projection A projection classification that maintains the shape and angles of a small area while minimizing distortion. As regions become larger, projections and area shapes become more distorted.

conic projection A type of map projection in which the globe is represented at an exact scale along a particular parallel (standard parallel) between the equator and a pole. Distortion increases away from that standard parallel. It can be envisioned that this map projection is made by projecting the features of the Earth onto a conic wrap (like a birthday hat) then cutting it to make a flat map.

continuous data A numeric form of data usually associated with the physical measurement of boundaries that are not well defined.

coordinate A number set that represents a specific location within a reference system. An *x-y* set is used in a planar coordinate system and an *x-y-z* set is used in a three-dimensional coordinate system.

curvilinear surface A surface consisting of or bounded by curved lines.

cylindric projection A type of map projection where the globe is represented by equidistant, parallel straight lines that intersect one another at right angles. The projection is true at the equator and increases in distortion toward the poles. It can be envisioned that this map projection is made by projecting the features of the Earth onto a cylindrical wrap then cutting it to make a flat map.

database management system (DBMS)	A program or collection of programs that enables the user to save, modify, classify, select, and extract information from a central database. DBMSs are used throughout the world and throughout almost every industry for the purposes of centralizing information and automating processes.
datum	Any set of numeric or geometric parameters used to accurately measure or define another quantity.
developable projection surfaces	A shape that can be flattened as a surface without any residual distortion. Predominant shapes include cones, cylinders, and planes.
digital elevation model (DEM)	Digital terrestrial or terrain elevation data that is typically defined by *z*-values and referenced to a common datum.
digital terrain model (DTM)	A digital, three-dimensional model of the Earth's surface. DTMs are typically used to define terrain and landform reliefs.
discrete data	Data about geographic features with well-defined boundaries that are represented by points, lines, or polygons.
distortion	The warping of shape, direction, scale, or area on a map projection or image in relation to real-world measurements.
distortion magnitude	The amount and density of distortion in a projection.
distortion pattern	A discernible pattern in a projection's distortion, such as the misrepresentation of landforms near the poles or distortion of areas.
dynamic pages	Web pages that offer some user interaction, whereby the pages display the most recent or dynamic database information as defined by the user.
Earth model	A physical shape that wholly represents the Earth and is used for mathematical and scientific purposes.
Earth-Observing Satellites (SPOT)	*See* Satellite Pour L'Observation de la Terre.
Eccentricity	An ellipse-specific constant that measures the degree of flattening as the relationship between the focal radius and the semimajor axis.

ellipse A two-dimensional shape that is flattened on one axis. The shape resembles a flattened circle and consists of a major and a minor axis.

ellipsoid A three-dimensional shape that is flattened on one axis. The shape resembles a flattened sphere and is formed from an ellipse rotated around an axis (axis of rotation). Like the ellipse, it consists of a major and minor axis.

ellipsoid of revolution An ellipsoid that revolves around a major axis. *See* axis of rotation

enterprise system A system that supports a wide variety of users having different functional requirements over an entire organization.

equal-area projection A projection classification determined when the relative sizes of all features on a globe are maintained during the projection process. An equal-area projection vigilantly retains area properties of the ellipsoid on the plane surface through the use of compensatory scale factors. This projection classification may distort shape, angle, and/or scale.

equator An imaginary line that covers the widest circumference on Earth. It is indicated as 0° latitude.

equatorial Relating to or dependent upon the equator.

equidistant projection A projection classification determined when scale is true along at least one line (a focal line) or from one or two points (focal points) to all other points on the projected surface.

Eratosthenes (276–194 B.C.) An innovative third-century B.C. Greek mathematician who made a nearly accurate calculation of the Earth's circumference using base scientific understanding, crude measuring instruments, and meticulous observation. His observations occurred in Egypt, within his home town of Syene (now Aswan) and at Alexandria, a town roughly 500 miles away.

feature file A file that contains geographic object feature information, such as representative point, line, and polygon information.

file-based system The simplest of relational database types that uses a single file for the database tables, queries, security information, and forms/reports.

finite Within a limit or set parameter.

first eccentricity *See* eccentricity.

flattening A measurement of axial compression in an ellipse that is directly related to the differences in both the semimajor and semiminor axes.

focal radius The distance from the center point to the foci (i.e., F_1, F_2) in an ellipse.

geocode Geospatially coded information products.

geodesy The division of science associated with the measurement and portrayal of the Earth. Geodesy, known also as geodetics, is intimately concerned with the determination of the size and shape of the Earth, as well as its elements. These Earth-based elements include its terrestrial gravity, magnetic field, tides, geologic and crustal movement, and polar motion.

geographic coordinate system A geographic reference system that uses latitudal and longitudal measurements to define point locations on a sphere or ellipsoid.

geographic data Geographic-related information about a map's features, including point locations, shapes, and descriptions.

geographic information *See* geographic data.

geographic information system (GIS) A collection of computer hardware, software, and geographic data for capturing, storing, updating, manipulating, analyzing, and displaying all forms of geographically referenced information. GIS is, in essence, a central repository of geographic data collected from various sources, including satellites, the global positioning system (GPS), and topographical maps.

geography The study of the Earth's surface that furnishes information about the Earth and distinguishes how features upon the Earth correlate with one another. It is the founding science to GISs.

geoid An exacting model of the Earth defined by the mean sea level. The model continues through the continents and is not defined by true topological heights or standard shape references.

geomensuration The measurement of the Earth as a whole, such as on a global level, that offers credible measurements useable toward precise geodetic positioning, surface shape, and size.

geometric A term that represents the spatial relationship between features and objects.

geometric deformation indicators The circular or elliptical indicators on an indicatrix. A circle represents conformality (no distortion in shapes), while an ellipse represents local distortion at the intersection where the object is plotted. These circular or elliptical objects are fixed at a finite scale with explicit locations in the projected surface.

geoprocessing The fundamental process of creating a derived set of geographic data from various existing data sets using operations such as feature overlay and data conversion.

georeferencing The process of referencing images, spatially transforming them into a defined coordinate referencing system, and integrating and overlaying them in a GIS environment.

geospatial analysis The process of analyzing and interpreting geospatial models.

global positioning system (GPS) Handheld and mobile devices that communicate with a system of satellites orbiting the Earth to record *x*, *y*, and *z* coordinates and calculate precise positional data. GPS devices are used for a multitude of applications, including surveying, mapping, hiking, and navigation.

globe A scaled, physical representation of the Earth.

graphic scale A map scale that depicts a bar of a certain length to represent a distance length. Graphic scales are commonly called bar scales and enable the map user to visually recognize distances with relative ease.

graticule A basic grid of meridians and parallels.

hemisphere One-half of a symmetrical shape, such as an ellipsoid or sphere. Earth is divided by its equator, forming a northern and a southern hemisphere.

horizontal control datum Any datum that serves as a geodetic reference for a precise northerly or southerly location (latitude) and easterly or westerly location (longitude).

index file A file that contains unique identifiers that comprise more detailed product information and help speed spatial feature queries.

indicatrix A visual distortion analysis tool that depicts a projected graticule as a distortion graph and shows distortion as circles or ellipses at the specific intersections in the plane.

inverse flattening The reciprocal of flattening used to minimize human error and help simplify conveyance.

irregular triangular surface model *See* triangulated irregular network.

isarithms *See* isoline.

isoline A line on any surface that connects points of equal value.

large-scale A project or map signifier that refers to applications involving small areas of focus, such as a community, a city block, or a university campus. A typical large-scale representative fraction can be assumed to be anything less than 1:50,000.

latitude The distance along a meridian that is either north or south of the equator. Latitudes are measured in degrees and are commonly called parallels.

layer A set of theme-based data that is described and stored in a map's library. Layers may contain a basemap or satellite image or can include such information as roadways, pipelines, soil sample locations, soil types, vegetation types, and water distribution.

legend A map reference that lists and distinguishes various map colors, lines, shapes, patterns, symbols, and annotation. Many times, the legend highlights map scale, source, projection, and other information.

light intensity detection and ranging (LIDAR) Remote sensing tool that uses a laser to precisely measure distances to various surfaces and terrain.

linked attribute table A matrixed table that contains explicative attributes for a group of spatial features.

longitude The angular distance between a target meridian and the prime meridian (0°). Longitude lines converge at the north and south poles, with measurements typically expressed in degrees, minutes, and seconds.

major axis A line of symmetry that passes through the widest portion of a shape. In an ellipse, the axis passes through both foci.

map projection A process that transforms a three-dimensional object, such as the Earth's globe, into a two-dimensional format. Some projections preserve shape; others preserve the accuracy of distance, area, or direction.

map scale The relationship between a distance on a map and its corresponding distance on the ground. It relates the size of map-based features with their actual, real-world size.

map space A platform for geographic data in depicting Earth features on a distinct coordinate system.

meridian A vertical line extending from the north pole to the south pole. Locations along the same meridian are said to have the same longitude.

metadata Information about a specific data set, including the geographic data's source, resolution, scale, and proper use.

mid-scale A project or map signifier that refers to applications that are not overly large or small, but rather fall right in the middle. A typical representative fraction ranges from 1:50,000 to 1:500,000.

minor axis A line of symmetry that passes through the shortest portion of a shape. In an ellipse, the minor axis passes through the center point and is perpendicular to the major axis.

multitier system A relational database type that functions similarly to the client-server system, except that there are middle servers, which contain software to perform certain tasks and process transmitted data. *See* three-tier system.

natural resources Natural elements of the Earth, such as rivers, streams, wetlands, and landforms.

natural scale *See* ratio scale.

Newton, Sir Isaac (1642–1727) Considered the father of modern mathematics, his gravitational theory published in *Philosophiae Naturalis Principia Mathematica* became the backbone for Britain's Earth model. The British model was an oblate ellipsoid—grapefruit-shaped and bulging in the middle. In 1745, the French Academy of Sciences proved the British model was correct and, subsequently, validated Newton's theory.

normal aspect A projection aspect based upon a standard line along which distortion is minimized. The farther you move away from the standard line, the greater the distortion.

oblate ellipsoid A grapefruit-shaped ellipsoid that bulges in the middle and is flat on top and bottom.

oblique aspect A projection aspect used when the axis of the Earth and the axis of the projection are oriented in an arbitrary manner. This aspect type is uniquely suited for geographic areas that are centered along lines that are neither meridians nor parallels. Consequently, this aspect type tends to be directed northwest, northeast, southwest, or southeast.

orthomorphic projection *See* conformal projection.

orthophotography The photographic process by which aerial photographs are rectified to produce an accurate image of the terrain. The process removes the Earth's tilt and any land relief displacements that occurred at the time the photo was taken.

parallel Two or more lines wholly separated by a constant distance. Geographically speaking, latitude lines, such as the equator, are in parallel with one another.

photogrammetry The process of identifying objects and measuring their true physical dimensions from photographs.

planar projection A type of map projection that is an azimuthal orientation of a projected surface. This projection type is true only at the center point and protects the integrity of azimuths, bearings, and directions from a central point to other more remote points in the plane. It can be envisioned that this map projection is made by projecting the features of the Earth onto a flat, touching plane, like a piece of flat paper touching a globe.

planar space A map space utilizing a horizontal (*x*) and vertical (*y*) axis.

planar surface A two-dimensional surface defined by an *x-y* coordinate grid.

plane coordinates Positional measurements from the origin (0,0) in a planar coordinate system. Coordinates are typically represented by an *x-y* pair.

plane of the ecliptic The imaginary line that defines the elliptical path the Earth takes as it orbits the Sun.

polar aspect A projection aspect that centers on a single point, either the north pole or the south pole. This aspect retains symmetry about any meridian and distorts between them.

poles The points where the axis of rotation intersects the surface of a three-dimensional shape.

primary key A unique attribute in a relational database management system (RDBMS) that distinguishes each record. They can be sequential record numbers or user-defined values that are unique and, thus, not duplicated anywhere within the same attribute throughout the records.

prime meridian A reference longitudal line that flows through Greenwich, England. The prime meridian is typically called the Greenwich meridian and is located at 0° east and west.

projection aspect The relative orientation of the projected plane and the ellipsoid of revolution with respect to the location of an observer.

projection type A term that relates to the distinct types of map projections available for use on the projected plane, such as the cylindric, conic, and planar types.

prolate ellipsoid An egg-shaped or elongated ellipsoid that is flattened in the middle.

quadrangle A map that defines a focal area's topography within latitude and longitude lines. It usually refers to the U.S. Geological Survey's 7.5-minute series or 15-minute series.

raster data Spatial data represented as a grid of valued pixels or cells. Each cell contains an attribute value and positional coordinates. These positional coordinates are wholly associated with the grid and its defined order.

ratio scale A map scale that reduces a map's distance to 1. One unit on the map corresponds to a distance in the same units on the ground. In a typical 30-minute series topographic map, the representative fraction is shown as either 1/125,000 or 1:125,000, whereby the 1 and 125,000 are the same unit of measure. A ratio scale is more formally known as a representative fraction and less formally as a natural scale.

reference ellipsoid A standardized ellipsoid with officially proved and measured parameters. The reference ellipsoid enables GIS to define point locations with regional accuracy.

region of interest The area where focus has been placed.

relational database management system (RDBMS) A program or collection of programs that enables the user to build data relationships to efficiently save, modify, classify, select, and extract information from a central relational database.

relative accuracy Map accuracy specific to an object's geographic location in relation to the true location of other objects on Earth.

relief map A map detailing the elevations and depressions of the Earth's surface with contour lines, shading, and digital terrain modeling. Relief maps are and appear to be three dimensional in nature.

remote sensing Any form of distance-based or remote viewing, observation, and measurement of the Earth. Typical forms include aerial photographs, satellite images, global positioning systems, LIDAR, and sonar.

representative fraction *See* ratio scale

resolution A measurement of image sharpness, usually indicated by pixels per line or dots per inch.

Satellite Pour L'Observation de la Terre (SPOT) A series of Earth-observing satellites that offer the ability to observe the whole Earth (minus a strip over the equator) in a single day. These satellites are part of a remote sensing program first designed by France in partnership with

	Belgium and Sweden. Current satellites include the SPOT1 and SPOT2 units. Use of the SPOT3 satellite ceased in 1996.
scale reduction	The process of proportionally reducing actual dimensions into much smaller, more convenient dimensions.
second eccentricity	An ellipse-specific constant that measures the degree of flattening as a relationship between the focal radius and the semiminor axis.
semimajor axis	Equal to one-half of the major axis.
semiminor axis	Equal to one-half of the minor axis.
sidereal day	A true day measurement, with a length of 23 hours, 56 minutes, and 4.09 seconds.
sidereal year	A true year measurement, with a length of 365.2564 days.
Six Sigma	Stands for six standard deviations from the mean or average and involves a measure of quality that strives for near perfection. This quality protocol is a disciplined, data-driven methodology for eliminating shortcomings and deficiencies in any business process.
small-scale	A project or map signifier that refers to applications involving relatively large areas, such as a map of the world, the continent of Australia, or the city of Rome, Italy. A typical small-scale representative fraction can be assumed to be anything greater than 1:500,000.
Snow, John (1813–1858)	A British physician who mapped all cholera deaths on a hand-drawn street map of a London neighborhood and helped end the epidemic of 1854.
spatial	A term used to describe anything located within a space that has physical, measurable dimensions.
spatial analysis	The study of geographic feature locations, attributes, and spatial dimensions and the relationships between each.
sphere	A symmetrical three-dimensional shape, in which the distance (radius) from the surface to a common center point is identical anywhere on the object. It can be thought as a circle rotated around an axis.
spheroid	*See* ellipsoid

standard axis *See* axis

standard line A line of exact scale that is represented by a true scale factor value of 1.

State Plane Coordinate System A two-dimensional, planar coordinate system established by the U.S. government to define positions on, above, and below the Earth's surface by *x-y* coordinates.

static pages Web pages that display database-driven information that cannot be changed, altered, or modified by the user.

Structured Query Language (SQL) A relational programming language used in complex database management systems. It is pronounced "Sequel."

Summer Solstice The longest day of the year, which occurs when the Sun is farthest north. In the northern hemisphere, the day is around June 21. In the southern hemisphere, the day is around December 22.

surveying The measurement of individual parts on the Earth's surface that offers credible measurements useable toward precise geodetic positioning of surface shape and size of the ever-changing physical terrain.

tagged-image file format (TIFF) A standard raster data file format that supports black-and-white, grayscale, limited color, and true color images.

thematic mapping The process of creating specialized maps using select information (i.e., a basemap) relating to one or more explicit themes. Themes include such information as roadways, pipelines, soil sample locations, soil types, vegetation types, and water distribution.

thick clients Users with a majority of processing capabilities in a relational database management system (RDBMS), such as is in the client-server and file-based relational database systems.

thin clients Users with reduced processing capabilities in a relational database management system (RDBMS), such as in the multitier relational database type.

three-dimensional Term pertaining to a spatial object or phenomenon dealing with three physical dimensions, typically *x*, *y*, and *z*.

three-dimensional mapping Spatial mapping that is detailed against three spatial coordinate planes, usually *x*, *y*, and *z*. Sometimes the *z* plane is substituted for an *m*, *n*, or k plane.

three-dimensional space A map space utilizing a horizontal (*x*), vertical (*y*), and perpendicular (z) axis.

three-tier system A relational database setup involving three distinct tiers: the client or user tier, the application tier, and the development tier. *See* multitier system.

Tissot indicatrix A helpful tool for visually modeling distortion and for graphically analyzing the distortion properties in a projected surface.

Tissot, Nicolas Auguste A nineteenth-century cartographer best known for his groundbreaking approach to distortion analysis, known now as the Tissot indicatrix, in his *Memoire sur la Representation des Surfaces et les Projections des Cartes Geographiques.* His indicatrix became a helpful tool for visually modeling distortion and for graphically analyzing the distortion properties in a projected surface.

Tobler's First Law of Geography A fundamental law of geography that states that things close together in space tend to be more alike than things that are far apart.

topographic map *See* quadrangle

topography A definition of the terrain's shape or configuration, typically represented in map form by layered contour lines.

topology The spatial relationships between contiguous or adjacent geographic features.

transformation The conversion of map or image coordinates from one coordinate system to another. Standard transformation techniques include rotation, skewing, shifting, and scaling.

transverse aspect The projection aspect when a standard line (normal) aspect is rotated 90°. This transverse aspect projection minimizes distortion along meridians and can have azimuthal or equatorial alignment.

triangulated irregular network (TIN) A network or series of continuous, nonoverlapping triangles used to define elevation data over a focus area. These triangles are used for surface representation and display, sometimes called an irregular triangular surface model.

true scale A scale that is devoid of distortion, represented by a scale factor of 1.

tuples Rows in a relational database management system (RDBMS) that contain the individual records.

two-dimensional Term pertaining to a spatial object or phenomenon dealing with only two physical dimensions, typically *x* and *y*.

two-dimensional space *See* planar space

UTM coordinate system A two-dimensional, positional coordinate system based upon the Universal Transverse Mercator (UTM) map projection. The UTM system partitions the earth's surface into 60 zones wherein each zone (cell) is roughly 6° east to west and 8° north to south.

vector data Spatial data expressed as an ordered list of coordinate points (*x-y* coordinates) to represent geographic features with associated attributes. Geographic features are represented by points, lines, and polygons.

vector space A platform for geographic vector data, which use *x-y* coordinates with lines and shapes to depict Earth features. It stores nontopological coordinate geometry and attribute information for spatial features.

verbal scale A map scale that expresses the map-to-ground relationship in words, such as "One inch represents 75 miles" or "One inch equals 75 miles."

vertical control datum Any datum that serves as a geodetic reference for elevations or height above and below the reference shape, be it a geoid or standard ellipsoid.

Winter Solstice The shortest day of the year, which occurs when the Sun is farthest south. In the northern hemisphere, the day is around December 22. In the southern hemisphere, the day is around June 21.

zenithal projection *See* planar projection

Bibliography

Advanced Information Systems, et al., *Oracle Unleashed*, 2nd ed., Indianapolis, IN: SAMS Publishing, 1997.

American National Standards Institute, *The Dublin Core Metadata Element Set,* ANSI/NISO Z39.85-2001, Bethesda, MD: NISO Press, 2001.

Angel, A. R., and S. R. Porter, *A Survey of Mathematics with Applications*, 6th ed., Reading, MA: Addison Wesley Longman, 2001.

Anton, H., *Calculus with Analytic Geometry*, 3rd ed., New York: John Wiley & Sons, 1988.

Audet, R., and G. Ludwig, *GIS in Schools*, Redlands, CA: ESRI Press, 2000.

Australian Drilling Industry Training Committee Ltd., *Drilling: The Manual of Methods, Applications, and Management*, 4th ed., A. W. Starkweather, Jr., (ed.), Boca Raton, FL: CRC Press—Lewis Publishers, 1997.

Australian Government Geoscience Australia, "Parameters for Commonly Used Ellipsoids," http://www.ga.gov.au/geodesy/datums/ellipsds.jsp (accessed June 27, 2005).

Baumann, G., and J. Weinstein, *Poised for Growth: Taking Your Business to the Next Level*, New York: McGraw-Hill, 1997.

Baumann, J., "GIS Helps Transportation Industry Keep on Truckin' in Czech Republic," *DirectionsMag.com*, http://www.directionsmag.com/article.php?article_id=532 (accessed May 2, 2005).

Belt, K., and S. T. Paxton, "GIS as an Aid to Visualizing and Mapping Geology and Rock Properties in Regions of Subtle Topography," *GSA Bull.*, Vol. 117, Nos. 1 and 2, 2005.

Berry, J. K., "Use 'Shadow Maps' to Understand Overlay Errors," *GeoWorld*, Vol. 17, No. 9, September 2004, pp. 18–19.

Board of Earth Sciences and Resources, *Down to Earth: Geographical Information for Sustainable Development in Africa*, Washington, D.C.: The National Academies Press, 2002.

Borkowski, K. M, "Accurate Algorithms to Transform Geocentric to Geodetic Coordinates," *Bull. Geod.,* Vol. 63, 1989, pp. 50–56.

Bothamley, J., *Dictionary of Theories*, London, U.K.: Gale Research International, 1993.

Boucher, C., et al., *The ITRF2000,* IERS Technical Note No. 31, Marne-la-Valle, France: International Earth Rotation and Reference Systems Service, 2004.

Brainerd, J., and A. Pang, *Floating Ring: A New Tool for Visualizing Distortion in Map Projections*, Santa Cruz, CA: Computer Science Department University of California, ftp://ftp.cse.ucsc.edu/pub/reinas/papers/cgi98.map.ps.Z.

Brainerd, J., and A. Pang, *Interactive Map Projections and Distortion*, Santa Cruz, CA: Computer Science Department University of California, ftp://ftp.cse.ucsc.edu/pub/reinas/papers/cgi98.map.ps.Z.

Bray, T., "RDF and Metadata," *XML.com*, August 1998, http://www.xml.com/pub/a/98/06/rdf.html (accessed November 10, 2005).

Brennan, P., "Dynamic Segmentation in ArcGIS," *ArcUser*, October–December 2002, pp. 52–54.

Briggs, R., "The Future of GIS: If I Really Knew, Would I Still Be Here?," University of Dallas, GISC 6381 GIS Fundamentals, 2004, http://www.utdallas.edu/~briggs/poec5319/future.ppt (accessed October 25, 2005).

Burke, R., *Getting to Know ArcObjects: Programming ArcGIS with VBA*, Redlands, CA: ESRI Press, 2003.

Bursa, M., "Dimension of the Earth's Mean Ellipsoid Specified by Brovar's Modified Conditions," *Acta Geod. Geoph. Hung.*, Vol. 39, No. 1, 2004, pp. 15–25.

Bursa, M., et al., "Dimension of the Earth's General Ellipsoid," *Earth, Moon, Planets*, Vol. 91, 2002, pp. 31–41.

Bush, G. W., "Executive Order 13286: Coordinating Geographic Data Acquisition and Access: The National Spatial Data Infrastructure (Amended)," *Federal Register*, Vol. 68, No. 43, March 5, 2003, pp. 10619–10633.

Butler, A., "Metadata Freedom: Exchange Data Without National Standards," *GeoWorld*, Vol. 18, No. 8, August 2005, pp. 36–39.

Butler, A., et al., "The National Geospatial Coordinating Council: A Dramatic New Approach to Build the NSDI," *GeoWorld*, Vol. 18, No. 10, October 2005, pp. 38–41.

Butte County Association of Governments, "GIS Enhances Transportation Planning," *BCAG Bulletin*, Fall 1997, p. 1.

Clarke, K. C., *Getting Started with Geographic Information Systems*, Upper Saddle River, NJ: Prentice Hall, 1997.

Clarke, K. C., S. L. McLafferty, and B. J. Tempalski, "On Epidemiology and Geographic Information Systems: A Review and Discussion of Future Directions," *Emerging Infectious Dis.*, Vol. 2, No. 2, April–June 1996, pp. 85–92.

Clinton, W. J., "Executive Order 12906: Coordinating Geographic Data Acquisition and Access: The National Spatial Data Infrastructure," *Federal Register*, Vol. 59, No. 71, April 13, 1994, pp. 17671–17674.

Clinton, W. J., "Executive Order 12951: Release of Imagery Acquired by Space-Based National Intelligence Reconnaissance Systems," *Federal Register*, Vol. 60, No. 39, February 28, 1995, pp. 10789–10790.

Cloud, J., "American Cartographic Transformations During the Cold War," *Cartography and Geographic Information Science*, Vol. 29, No. 3, 2002, pp. 261–282.

Codd, E. F., "A Relational Model of Data for Large Shared Data Banks," *Commun. ACM*, Vol. 13, No. 6, June 1970, pp. 377–387.

Coddington, W., *Environmental Marketing: Positive Strategies for Reaching the Green Consumer*, New York: McGraw-Hill, 1993.

Cova, T. J., and R. L. Church, "Modeling Community Evacuation Vulnerability Using GIS," *Int. J. Geographical Information Sci.*, Vol. 11, No. 8, 1997, pp. 763–784.

Crossley, R., "Future of GIS," *GeoCommunity,* http://spatialnews.geocomm.com/features/futuregis.html (accessed November 5, 2005).

Dana, P. H., "Geodetic Datum List," *The Geographer's Craft Project*, Department of Geography, University of Texas–Austin, July 12, 1997, http://www.colorado.edu/geography/gcraft/notes/datum/dlist.html (accessed June 23, 2005).

Dana, P. H., "Reference Ellipsoids and Geodetic Datum Transformation Parameters (Local to WGS-84) from NIMA 8350.2 4 July 1977 and MADTRAN 1 October 1996," http://www.colorado.edu/geography/gcraft/notes/datum/edlist.html (accessed June 23, 2005).

Daniels, T., and K. Chhabra, "A Mapping Marriage: Cartography and Mapping Forge a Beneficial Union," *GeoWorld*, Vol. 18, No. 10, October 2005, pp. 34–37.

Davis, B. E., *GIS: A Visual Approach*, Santa Fe, NM: OnWord Press, 1996.

Davis, B. E., *GIS for Everyone,* 3rd ed., Redlands, CA: ESRI Press, 2003.

Defense Mapping Agency, "Datums, Ellipsoids, Grids, and Grid Reference Systems," *Defense Mapping Agency Technical Manual 8358.1*, Washington, D.C.: Defense Mapping Agency, 1990.

Defense Mapping Agency, "Error Theory as Applied to Mapping, Charting, and Geodesy," *Defense Mapping Agency Technical Manual 8400.1*, Washington, D.C.: Defense Mapping Agency, 1991.

Defense Mapping Agency, Geodesy for the Layman, Defense Mapping Agency Technical Report 80-D03, Washington, D.C.: Defense Mapping Agency, 1984.

Defense Mapping Agency, "The Universal Grids: Universal Transverse Mercator (UTM) and Universal Polar Stereographic (UPS)," *Defense Mapping Agency Technical Manual 8358.2*, Washington, D.C.: Defense Mapping Agency, 1989.

Defense Mapping School, Department of Defense, *Basic Geodesy,* Student Pamphlet 4MF1-F-010-010, Washington, D.C.: National Oceanic and Atmospheric Administration, 1977.

Dewhurst, W. T., *NADCON: The Application of Minimum Curvature-Derived Surfaces in the Transformation of Positional Data from the North American Datum of 1927 to the North American Datum of 1983,* NOAA Technical Report NOS NG-50, Rockville, MD: National Oceanic and Atmospheric Administration, 1990.

DiBiase, D., "Lesson 2: Scales and Projections," Geography 5121 Course, Pennsylvania State University, 2004.

DiBiase, D., "Lesson 3: Census Data and Thematic Maps," Geography 5121 Course, Pennsylvania State University, 2004.

Digital Geographic Information Working Group, *The Digital Geographic Information Exchange Standard (DIGEST): Part 3 Codes and Parameters*, 2.1 ed., September 2000.

Downing, D., *Dictionary of Mathematics Terms*, 2nd ed., Hauppauge, NY: Barron's Educational Series, 1995.

Dracup, J. F., *Preprint—Understanding the State Plane Coordinate Systems,* Rockville, MD: U.S. Department of Commerce, National Oceanic and Atmospheric Administration, National Ocean Service, 1994.

Dracup, J. F., *Horizontal Control,* NOAA Technical Report NOS 88 NGS 19, Rockville, MD: National Oceanic and Atmospheric Administration, 1980.

Environmental Systems Research Institute, "About Topology," ArcGIS Desktop Help, *ArcMap 9.0 Build 560,* Redlands, CA: ESRI Press, 2004.

Environmental Systems Research Institute, *ArcIMS 4 Architecture and Functionality*, ESRI White Paper, Redlands, CA: ESRI Press, May 2003.

Environmental Systems Research Institute, *ArcMap 9.0 Build 560,* Redlands, CA: ESRI Press, 2004.

Environmental Systems Research Institute, *Dictionary of GIS Terminology*, H. Kennedy, (ed.), Redlands, CA: ESRI Press, 2001.

Environmental Systems Research Institute, "Environmental Site Assessment," *Environmental Observer*, Winter 2005, pp. 4–5.

Environmental Systems Research Institute, *ESRI Map Book Volume 16–Geography–Creating Communities*, Redlands, CA: ESRI Press, 2001.

Environmental Systems Research Institute, *ESRI Map Book Volume 17–Geography and GIS–Sustaining Our World*, Redlands, CA: ESRI Press, 2002.

Environmental Systems Research Institute, *ESRI Map Book Volume 18–Geography and GIS–Serving Our World*, Redlands, CA: ESRI Press, 2003.

Environmental Systems Research Institute, *ESRI Shapefile Technical Description*, ESRI White Paper, Redlands, CA: ESRI Press, July 1998.

Environmental Systems Research Institute, *Metadata and GIS*, ESRI White Paper, Redlands, CA: ESRI Press, October 2002.

Environmental Systems Research Institute, "To Monitor and Inspect Underground Storage Tanks, the Illinois Office of the State Fire Marshal Uses GIS," *ArcNews*, Vol. 25, No. 3, Fall 2003, pp. 15.

Environmental Systems Research Institute, "Topology Rules," ArcGIS Desktop Help, *ArcMap 9.0 Build 560,* Redlands, CA: ESRI Press, 2004.

European Space Agency, "Benefits for Europe," http://www.esa.int/esaCP/SEMO75XO4HD_Benefits_0.html (accessed February 7, 2005).

Executive Office of the President, Office of Management and Budget, *Circular A-16 Revised*, Washington, D.C.: Executive Office of the President, 2002.

Federal Energy Regulatory Commission, *Guidance for Shoreline Management Planning at Hydropower Facilities*, Washington, D.C.: FERC, 2001.

Federal Geographic Data Committee, *Content Standard for Digital Geospatial Metadata Workbook*, Ver. 2.0, Washington, D.C.: FGDC, 2000.

Federal Geographic Data Committee, "FGDC Don't Duck Metadata: Metadata Quick Guide," Ver. March 2005, http://www.fgdc.gov/metadata/education/MetadataQuickGuide.pdf (accessed September 15, 2005).

Federal Geographic Data Committee, *Geospatial Positioning Accuracy Standards—Part 1: Reporting Methodology,* FGDC-STD-007.1-1998, Reston, VA: FGDC, 1998.

Federal Geographic Data Committee, Metadata Ad Hoc Working Group, *Content Standard for Digital Geospatial Metadata,* FGDC-STD-001-1998, Washington, D.C.: FGDC, 1998.

Galison, P., *Einstein's Clocks, Poincaré's Maps: Empires of Time,* New York: W. W. Norton & Company, 2003.

GIS Portal.com, "GIS Data and Software Library," http://www.gisportal.com/gis3f.htm (accessed October 5, 2005).

GIS.com, "What Is GIS?" http://www.gis.com/whatisgis/whatisgis.html (accessed November 15, 2004).

GISLounge.com, "Data and Resources Links," http://gislounge.com/ll/data.shtml (accessed October 1, 2005).

GlobalSecurity.org, "Appendix E: Ellipsoids and Datums," http://www.globalsecurity.org/military/library/policy/army/fm/6-2/Appe.htm#tabe_23 (accessed June 27, 2005).

Goodchild, M. F., *The Fundamental Laws of GIScience,* University of California Santa Barbara, http://www.csiss.org/aboutus/presentations/files/goodchild_ucgis_jun03.pdf (accessed April 5, 2005).

Goodchild, M. F., and D. M. Mark, "The Fractal Nature of Geographic Phenomena," *Ann. Assoc. Am. Geographers,* Vol. 77, No. 2, 1987, pp. 265–278.

Greene, R. W., *Confronting Catastrophe: A GIS Handbook,* Redlands, CA: ESRI Press, 2002.

Greene, R. W., *GIS in Public Policy: Using Geographic Information for More Effective Government,* Redlands, CA: ESRI Press, 2000.

Grise, S., and B. Brodaric, *ArcGIS Geology Data Model,* Redlands, CA: ESRI Press, 2004.

Haagmans, R., et al., "A Procedure for Combining Gravimetric Geoid Models and Independent Geoid Data, with an Example in the North Sea Region," *DEOS Prog. Lett.,* Vol. 98, No. 1, pp. 89–99.

Hammond, A. D., "An Application of GIS in Underground Mining," *Earth Observation Magazine,* Vol. 10, No. 9, September 2001, pp. 13–16.

Hart, D., and H. Phillips, "Metadata Primer: A 'How To' Guide on Metadata Implementation," *National States Geographic Information Council,* http://www.lic.wisc.edu/metadata/metaprim.htm (accessed August 12, 2005).

Heiskanen, W. A., and F. A. Vening Meinesz, *The Earth and Its Gravity Field,* New York: McGraw Hill, 1958.

Hightower, A. W., and R. E. Klein, "Building a Geographic Information System (GIS) Public Health Infrastructure for Research and Control of Tropical Diseases," *Emerging Infectious Dis.,* Vol. 1, No. 4, October–December 1995, pp. 156–157.

Hirsch, S., "19thCent_R.RTF (Nineteenth Century Map Projections in RTF)," *Map Projections and Coordinate Systems,* Department of Geography and Earth Systems Science 661, George Mason University, Fall Semester 1999, http://classweb.gmu.edu/shirsch/Map Projections/ (accessed September 27, 2004).

Huff, D. L., "Parameter Estimation in the Huff Model," *ArcUser,* October–December 2003, pp. 34–36.

Hunter Department of Geology & Geography, "Map Projection Distortion," http://www.geography.hunter.cuny.edu/mp/distort.html (accessed August 27, 2004).

Inter-Governmental Committee on Surveying and Mapping, *Standards and Practices for Control Surveys (SP1)*, ICSM Publication No. 1, Ver. 1.5, Australia: Geocentric Datum of Australia, 2002.

Intergraph Mapping and Geospatial Solutions, "Case Study," http://imgs.intergraph.com/ (accessed February 7, 2005).

International Organization for Standardization, Technical Committee 46/Subcommittee 4, *Information and Documentation—The Dublin Core Metadata Element Set,* ISO 15836:2003(E), Geneva, Switzerland: International Organization for Standardization, 2003.

Ireson, W. G., et al., *Handbook of Reliability Engineering and Management*, 2nd ed., New York: McGraw-Hill, 1996.

Jafrullah, M., et al., "An Integrated Approach for Banking GIS," GISdevelopment.net, http://www.gisdevelopment.net/application/business/mi03140pf.htm (accessed June 3, 2005).

Jo, M.-H., et al., "The Management System Development of Campus Facility Information Using Web-Based GIS," *ACRS 2000—Proc. 21st Asian Conf. Remote Sensing*, Taipei, Taiwan, December 4–8, 2000.

Johnston, K., *Using ArcGIS Spatial Analyst*, Redlands, CA: ESRI Press, 2002.

Jones, A., and G. Blick, *Where in the World Are We?*, Ver. 1.0, Wellington, New Zealand: Government of South Australia, 1999.

Kubal, M. T., et al., *Building Profits in the Construction Industry*, New York: McGraw-Hill, 2000.

Kulik, L., "Reasoning About Cardinal Directions Using Grids as Qualitative Geographic Coordinates," *COSIT'99, Lecture Notes in Computer Science,* D. M. M. Freksa, (ed.), Vol. 1661, 1999, pp. 205–220.

Lang, L., *Managing Natural Resources with GIS*, Redlands, CA: ESRI Press, 2003.

Laton, W. R., "Guide to Map Scale," California State University at Fullerton, Department of Geological Sciences, http://geology.fullerton.edu/Faculty/wlaton/Map-scale.htm (accessed June 30, 2005).

Letham, L., *GPS Made Easy*, 3rd ed., Seattle, WA: The Mountaineers, 2001.

Liebschultz, H. D., *The Oracle Cookbook: For Design, Administration, and Implementation*, New York: M&T Books, 1996.

Longley, P. A., et al., *Geographic Information Systems and Science*, New York: John Wiley & Sons, 2001.

MacEachren, A. M. *How Maps Work: Representation, Visualization, and Design*, 3rd ed., New York: Guilford Press, 1995.

Manning, J., and J. Steed, "Positional Accuracy," *Int. Symp. Spatial Data Infrastructure*, University of Melbourne, Australia, November 19–20, 2001.

Maryland–National Capital Park and Planning Commission, *Geographic Information System: Cost/Benefit Assessment Report M-NCPPC*, submitted to the Montgomery County Council's Management and Fiscal Policy Committee, April 12, 1999.

McCoy, J., *Geoprocessing in ArcGIS 9*, ESRI Software documentation, http://support.esri.com (accessed December 7, 2004).

McManus, B., "Thematic Mapping," GIS Resource Document 05-70, Geographic Information Analysis Core of the Population Research Institute, Pennsylvania State University, April 2005, http://www.pop.psu.edu/gia-core/pdfs/gis_rd_05-70.pdf (accessed November 20, 2005).

Mendelson, B., *Introduction to Topology*, 3rd ed., New York: Dover Publications, 1990.

Merrigan, M. J., et al., "A Refinement to the World Geodetic System 1984 Reference Frame," *Proc. ION-GPS-2002*, Institute of Navigation, Portland, OR, September 24–27, 2002, pp. 1519–1529.

Mitchell, R. L., "Web Services Put GIS on the Map," *Computerworld*, December 15, 2003.

Miugglier, C. J., "The French Republic," *Photogrammetric Eng. Remote Sensing*, January 2001, pp 33–35.

Molodensky, S. M., *Methods for the Study of the External Gravitational Field and Figure of the Earth*, New York: Ministry of Geology and Control of Mineral Resources of the USSR, Israel Program for Scientific Translations, 1962.

Molodensky, S. M., "On the Models of the Lower Mantle Viscosity Consistent with the Modern Data of Core-Mantle Boundary Flattening," *Stud. Geophys. Geod.*, Vol. 46, 2002, pp. 411–433.

Monmonier, M., *Air Apparent: How Meteorologists Learned to Map, Predict, and Dramatize Weather*, Chicago, IL: University of Chicago Press, 2000.

Monmonier, M., *How to Lie with Maps*, 2nd ed., Chicago, IL: University of Chicago Press, 1996.

Moore, L., *Transverse Mercator Projections and U.S. Geological Survey Digital Products*, Washington, DC: Mid-Continent Mapping Center, U.S. Geological Survey, 1997.

Morales, M., "GIS Software Components: Example Using OpenMap and MapObjects," *GISLounge*, http://gislounge.com/ucon/ucopenmap.shtml (accessed September 29, 2005).

Mueller, I. I., and J. D. Rockie, *Gravimetric and Celestial Geodesy*, New York: Frederick Ungar Publishing Co., 1966.

National Archives of Australia, *AGLS Metadata Element Set, Part 1: Reference Description*, Ver. 1.3, Commonwealth of Australia: National Archives of Australia, 2002.

National Geospatial Intelligence Agency, "Africa Datums Table," http://earth-info.nga.mil/GandG/coordsys/onlinedatum/CountryAfricaTable.html (accessed June 25, 2005).

National Geospatial Intelligence Agency, "Asia Datums Table," http://earth-info.nga.mil/GandG/coordsys/onlinedatum/CountryAsiaTable.html (accessed June 25, 2005).

National Geospatial Intelligence Agency, "Atlantic Ocean Datums Table," http://earth-info.nga.mil/GandG/coordsys/onlinedatum/CountryAtlanticTable.html (accessed June 25, 2005).

National Geospatial Intelligence Agency, "Australia Datums Table," http://earth-info.nga.mil/GandG/coordsys/onlinedatum/CountryAustraliaTable.html (accessed June 25, 2005).

National Geospatial Intelligence Agency, *Department of Defense World Geodetic System 1984: Its Definition and Relationships with Local Geodetic Systems*, 3rd ed., NIMA Stock No. TR8350.2, Bethesda, MD: NGIA, January 3, 2000.

National Geospatial Intelligence Agency, "Europe Datums Table," http://earth-info.nga.mil/GandG/coordsys/onlinedatum/CountryEuropeTable.html (accessed June 25, 2005).

National Geospatial Intelligence Agency, "Indian Ocean Datums Table," http://earth-info.nga.mil/GandG/coordsys/onlinedatum/CountryIndianTable.html (accessed June 25, 2005).

National Geospatial Intelligence Agency, "North America Datums Table," http://earth-info.nga.mil/GandG/coordsys/onlinedatum/CountryNorthAmericaTable.html (accessed June 25, 2005).

National Geospatial Intelligence Agency, "Pacific Ocean Datums Table," http://earth-info.nga.mil/GandG/coordsys/onlinedatum/CountryPacificTable.html (accessed June 25, 2005).

National Geospatial Intelligence Agency, "South American Datums Table," http://earth-info.nga.mil/GandG/coordsys/onlinedatum/CountrySouthAmericanTable.html (accessed June 25, 2005).

National Information Standards Organization, *Understanding Metadata,* Bethesda, MD: NISO Press, 2004.

Natural Resources Canada, Geodetic Survey Division, "Geodesy," http://www.geod.nrcan.gc.ca/index_e/geodesy_e/whatis_e.html (accessed May 18, 2005).

Noobeed.com, "List of Predefined Datums," http://www.noobeed.com/nb_ex_geod_datum_list.htm (accessed June 25, 2005).

Oberlin College, "Thematic Mapping: Some Basics," http://www.oberlin.edu/OCTET/HowTo/GIS/Tutorial/3_ThematicMapping/Thematic%20Maps.htm (accessed November 17, 2005).

Open GIS Consortium, "OpenGIS Implementation Specification: Coordinate Transformation Services," *Document OGC 01-009,* Rev. 1.00, Wayland, MA: Open GIS Consortium, 2001.

Open GIS Consortium, "OpenGIS Reference Model," Document OGC 03-040, Kurt Buehler, (ed.), Ver. 0.1.2, Wayland, MA: Open GIS Consortium, 2003.

Open GIS Consortium, *OpenGIS® Simple Features Specification for SQL,* Rev. 1.1, Project Document 99-049, Wayland, MA: Open GIS Consortium, May 5, 1999.

Open GIS Consortium, "Topic 11: OpenGIS™ Metadata (ISO/TC 211 DIS 19115)," *The OpenGISTM Abstract Specification,* Ver. 5, Wayland, MA: Open GIS Consortium, 2001.

Papathoma, M., et al., "Assessing Tsunami Vulnerability: An Example from Herakleio, Crete," *Nat. Hazards Earth Syst. Sci.*, Vol. 3, 2003, pp. 377–389.

Price, M., "Incident Command and GIS," *ArcUser,* Vol. 7, No. 4, October–December 2004, pp. 34–37.

Raina, S., and V. Amanna, "GIS for Rural Marketing," *Proc. 6th Ann. Int. Map India Conf. 2003,* New Delhi, India, January 28–31, 2003.

Ratcliffe, J. H., "Geocoding Crime and a First Estimate of a Minimum Acceptable Hit Rate," *Int. J. Geographical Information Sci.*, Vol. 18, No. 1, January–February 2004, pp. 61–72.

Richards, T. B., et al., "Geographic Information Systems and Public Health: Mapping the Future," *Public Health Reports*, Vol. 114, 1999, pp. 359–373.

Riordan, R. M., *Designing Relational Database Systems*, Redmond, WA: Microsoft Press, 1999.

Ritchie, D., *C Reference Manual*, Murray Hill, NJ: Bell Telephone Laboratories, http://cm.bell-labs.com/cm/cs/who/dmr/cman.pdf (accessed February 4, 2005).

Robinson, A. H., et al., *Elements of Cartography*, 6th ed., New York: Wiley & Sons, 1995.

Ryan, B. J., et al., "A Clear Vision of the NSDI," *Geospatial Solutions*, April 1, 2004.

Ryburn, R. J., "The Evolution of Geoscientific Metadata," *AGSO Research Newsletter*, Vol. 31, November 1999, pp. 12–13.

Schweitzer, P. N., "Metadata in Plain Language," *US Geologic Survey, Formal Metadata: Techniques and Software*, http://geology.usgs.gov/tools/metadata/tools/doc/ctc/ (accessed August 12, 2005).

Schweitzer, P. N., *National Spatial Data Clearinghouse: The Real Story*, U.S. Geological Survey Open-File Report 03-471, Digital Mapping Techniques 2003 Workshop Proceedings, 2003.

Silva, E., *Cost–Benefit Analysis for Geographic Information System: Implementation Justification*, submitted to Bruce Oswald, Chair of the NYS GIS Coordinating Body, March 4, 1998.

Slocum, T. A., et al., *Thematic Cartography and Geographic Visualization*, 2nd ed., Upper Saddle River, NJ: Prentice Hall, 2003.

Smith, J. R., *The Meridian Arc Measurement in Peru 1735–1745*, International Institution for the History of Surveying and Measurement, http://www.ddl.org/figtree/hsm/index.htm (accessed August 14, 2005).

Snay, R., and T. Soler, "Modern Terrestrial Reference Systems Part 2: The Evolution of NAD83," *Professional Surveyor*, February 2000.

Snay, R., and T. Soler, "Modern Terrestrial Reference Systems Part 3: WGS84 to ITRS," *Professional Surveyor*, March 2000.

Snowdon, N., *Oracle Programming with Visual Basic*, San Francisco, CA: Sybex, 1999.

Snyder, J. P., *Map Projections: A Reference Manual*, London, U.K.: Taylor and Francis, 1995.

Snyder, J. P., *Map Projections: A Working Manual*, Washington, D.C.: U.S. Government Printing Office, 1987.

Snyder, J. P., and P. M. Voxland, "An Album of Map Projections," U.S. Geological Survey Professional Paper 1453, Washington, D.C.: U.S. Government Printing Office, 1989.

Sonoma State University, Geography and Global Studies, "Map Scale," http://www.sonoma.edu/geography/resources/mapscale.htm (accessed July 2, 2005).

Strahler, A. N., *Introduction to Physical Geography*, 3rd ed., New York: John Wiley & Sons, 1973.

Switala, K., "Crime Stoppers: Web-Based Mapping Makes Safer Neighborhoods," *Geoworld*, Vol. 18, No. 4, April 2005, pp. 36–38.

Tang, W., and J. Selwood, *Connecting Our World: GIS Web Services*, Redlands, CA: ESRI Press, 2003.

Tenzer, R., "Discussion of Mean Gravity Along the Plumbline," *Stud. Geophys. Geod.*, Vol. 48, 2004, pp. 309–330.

Tissot, A., *Mémoire sur la Représentation des Surfaces et les Projections des Cartes Géographiques,* Paris, France: Gauthier-Villars, 1881.

Tomlinson, R. F., *Thinking About GIS: Geographic Information System Planning for Managers*, Redlands, CA: ESRI Press, 2003.

Tracy, J., *Plane Surveying*, New York: John Wiley & Sons, 1906.

University Archives, University Libraries, and State University of New York at Buffalo, "Introduction to the Love Canal," *Love Canal Collection*, http://ublib.buffalo.edu/libraries/projects/lovecanal/introduction.html (accessed April 2, 2005).

U.S. Centers for Disease Control and Prevention, "John Snow," http://www.cdc.gov/ncidod/dbmd/snowinfo.htm (accessed February 7, 2005).

U.S. Environmental Protection Agency, *Indoor Air Pollution: An Introduction for Health Professionals*, Washington, D.C.: EPA, 2002.

U.S. Environmental Protection Agency, *Integrated Environmental Management Systems: Implementation Guide*, EPA 744-R-00-011, Washington, D.C.: EPA, October 2000.

Weiss, A., *Million Dollar Consulting: The Professional's Guide to Growing a Practice*, 2nd ed., New York: McGraw-Hill, 1992.

Wilkins, V., and L. Lundquist, "Innovations at Work: Rail," *MassTransit*, Vol. 31, No. 2, April 2005, pp. 40–44.

Xu, Y., and L. Liu, "GIS Based Analysis of Store Closure: A Case Study of an Office Depot Store in Cincinnati," *Geoinformatics 2004: Proc. 12th Int. Conf. Geoinformatics—Geospatial Information Research: Bridging the Pacific and Atlantic*, University of Gavle, Sweden, June 7–9, 2004.

Yiannakoulias, N., et al., "Zones of Prevention: The Geography of Fall Injuries in the Elderly," *Social Sci. Med.*, Vol. 57, 2003, pp. 2065–2073.

Yurkina, M. I., "To Pizzetti's Theory of the Level Ellipsoid," *Stud. Geophys. Geod.*, Vol. 46, 2002, pp. 435–454.

Zeiler, M., *Modeling Our World: The ESRI Guide to Geodatabase Design*, Redlands, CA: ESRI Press, 1999.

Zhao, L., "Integrating Rank Correlation Techniques with GIS for Marketing Analysis," *GeoComputation 2000, Proc. 5th Int. Conf. GeoComputation*, University of Greenwich's School of Earth and Environmental Sciences, Chatham Maritime, Kent, U.K., August 23–25, 2000.

Zilkoski, D. B., et al., *Guidelines for Establishing GPS-Derived Ellipsoid Heights (Standards: 2 cm and 5 cm),* NOAA Technical Memorandum NOS NG-58, Ver. 4.3, Silver Spring, MD: National Oceanic and Atmospheric Administration, 1987.

Zurayk, R., "Participatory GIS-Based Natural Resource Management: Experiences from a Country of the South," *Arid Lands Newsletter*, No. 53, May–June 2003.

About the Author

Stephen R. Galati is the Northeast proposal manager for TRC Environmental Corporation. With more than 13 years of experience in technical writing, engineering, and proposal management, Mr. Galati's work centers upon environmental, marketing, and GIS applications throughout the United States. He has written numerous articles for industry periodicals.

Mr. Galati holds a B.S. in electrical engineering from Pratt Institute and an M.A. in English from the College of Staten Island–City University of New York. He is a member of the American Society of Photogrammetry and Remote Sensing, the American Geophysical Union, and the New York City Chapter of Geographic Information Systems and Mapping Operations (GISMO), as well as a GeoData Alliance Affiliate.

Index

Recent Titles in the Artech House Mobile Communications Series

John Walker, Series Editor

3G CDMA2000 Wireless System Engineering, Samuel C. Yang

3G Multimedia Network Services, Accounting, and User Profiles, Freddy Ghys, Marcel Mampaey, Michel Smouts, and Arto Vaaraniemi

802.11 WLANs and IP Networking: Security, QoS, and Mobility, Anand R. Prasad, Neeli R. Prasad

Advances in 3G Enhanced Technologies for Wireless Communications, Jiangzhou Wang and Tung-Sang Ng, editors

Advances in Mobile Information Systems, John Walker, editor

Advances in Mobile Radio Access Networks, Y. Jay Guo

Applied Satellite Navigation Using GPS, GALILEO, and Augmentation Systems, Ramjee Prasad and Marina Ruggieri

CDMA for Wireless Personal Communications, Ramjee Prasad

CDMA Mobile Radio Design, John B. Groe and Lawrence E. Larson

CDMA RF System Engineering, Samuel C. Yang

CDMA Systems Capacity Engineering, Kiseon Kim and Insoo Koo

CDMA Systems Engineering Handbook, Jhong S. Lee and Leonard E. Miller

Cell Planning for Wireless Communications, Manuel F. Cátedra and Jesús Pérez-Arriaga

Cellular Communications: Worldwide Market Development, Garry A. Garrard

Cellular Mobile Systems Engineering, Saleh Faruque

The Complete Wireless Communications Professional: A Guide for Engineers and Managers, William Webb

EDGE for Mobile Internet, Emmanuel Seurre, Patrick Savelli, and Pierre-Jean Pietri

Emerging Public Safety Wireless Communication Systems, Robert I. Desourdis, Jr., et al.

The Future of Wireless Communications, William Webb

Geographic Information Systems Demystified, Stephen R. Galati

GPRS for Mobile Internet, Emmanuel Seurre, Patrick Savelli, and Pierre-Jean Pietri

GPRS: Gateway to Third Generation Mobile Networks, Gunnar Heine and Holger Sagkob

GSM and Personal Communications Handbook, Siegmund M. Redl, Matthias K. Weber, and Malcolm W. Oliphant

GSM Networks: Protocols, Terminology, and Implementation, Gunnar Heine

GSM System Engineering, Asha Mehrotra

Handbook of Land-Mobile Radio System Coverage, Garry C. Hess

Handbook of Mobile Radio Networks, Sami Tabbane

High-Speed Wireless ATM and LANs, Benny Bing

Interference Analysis and Reduction for Wireless Systems, Peter Stavroulakis

Introduction to 3G Mobile Communications, Second Edition, Juha Korhonen

Introduction to Communication Systems Simulation, Maurice Schiff

Introduction to Digital Professional Mobile Radio, Hans-Peter A. Ketterling

Introduction to GPS: The Global Positioning System, Ahmed El-Rabbany

An Introduction to GSM, Siegmund M. Redl, Matthias K. Weber, and Malcolm W. Oliphant

Introduction to Mobile Communications Engineering, José M. Hernando and F. Pérez-Fontán

Introduction to Radio Propagation for Fixed and Mobile Communications, John Doble

Introduction to Wireless Local Loop, Second Edition: Broadband and Narrowband Systems, William Webb

IS-136 TDMA Technology, Economics, and Services, Lawrence Harte, Adrian Smith, and Charles A. Jacobs

Location Management and Routing in Mobile Wireless Networks, Amitava Mukherjee, Somprakash Bandyopadhyay, and Debashis Saha

Mobile Data Communications Systems, Peter Wong and David Britland

Mobile IP Technology for M-Business, Mark Norris

Mobile Satellite Communications, Shingo Ohmori, Hiromitsu Wakana, and Seiichiro Kawase

Mobile Telecommunications Standards: GSM, UMTS, TETRA, and ERMES, Rudi Bekkers

Mobile Telecommunications: Standards, Regulation, and Applications, Rudi Bekkers and Jan Smits

Multiantenna Digital Radio Transmission, Massimiliano "Max" Martone

Multiantenna Wireless Communicatio ns Systems, Sergio Barbarossa

Multipath Phenomena in Cellular Networks, Nathan Blaunstein and Jørgen Bach Andersen

Multiuser Detection in CDMA Mobile Terminals, Piero Castoldi

OFDMA for Broadband Wireless Access, Sławomir Pietrzyk

Personal Wireless Communication with DECT and PWT, John Phillips and Gerard Mac Namee

Practical Wireless Data Modem Design, Jonathon Y. C. Cheah

Prime Codes with Applications to CDMA Optical and Wireless Networks, Guu-Chang Yang and Wing C. Kwong

QoS in Integrated 3G Networks, Robert Lloyd-Evans

Radio Engineering for Wireless Communication and Sensor Applications, Antti V. Räisänen and Arto Lehto

Radio Propagation in Cellular Networks, Nathan Blaunstein

Radio Resource Management for Wireless Networks, Jens Zander and Seong-Lyun Kim

RDS: The Radio Data System, Dietmar Kopitz and Bev Marks

Resource Allocation in Hierarchical Cellular Systems, Lauro Ortigoza-Guerrero and A. Hamid Aghvami

RF and Baseband Techniques for Software-Defined Radio Peter B. Kenington

RF and Microwave Circuit Design for Wireless Communications, Lawrence E. Larson, editor

Sample Rate Conversion in Software Configurable Radios, Tim Hentschel

Signal Processing Applications in CDMA Communications, Hui Liu

Smart Antenna Engineering, Ahmed El Zooghby

Software Defined Radio for 3G, Paul Burns

Spread Spectrum CDMA Systems for Wireless Communications, Savo G. Glisic and Branka Vucetic

Third Generation Wireless Systems, Volume 1: Post-Shannon Signal Architectures, George M. Calhoun

Traffic Analysis and Design of Wireless IP Networks, Toni Janevski

Transmission Systems Design Handbook for Wireless Networks, Harvey Lehpamer

UMTS and Mobile Computing, Alexander Joseph Huber and Josef Franz Huber

Understanding Cellular Radio, William Webb

Understanding Digital PCS: The TDMA Standard, Cameron Kelly Coursey

Understanding GPS: Principles and Applications, Second Edtion, Elliott D. Kaplan and Christopher J. Hegarty, editors

Understanding WAP: Wireless Applications, Devices, and Services, Marcel van der Heijden and Marcus Taylor, editors

Universal Wireless Personal Communications, Ramjee Prasad

WCDMA: Towards IP Mobility and Mobile Internet, Tero Ojanperä and Ramjee Prasad, editors

Wireless Communications in Developing Countries: Cellular and Satellite Systems, Rachael E. Schwartz

Wireless Intelligent Networking, Gerry Christensen, Paul G. Florack, and Robert Duncan

Wireless LAN Standards and Applications, Asunción Santamaría and Francisco J. López-Hernández, editors

Wireless Technician's Handbook, Second Edition, Andrew Miceli

For further information on these and other Artech House titles, including previously considered out-of-print books now available through our In-Print-Forever® (IPF®) program, contact:

Artech House
685 Canton Street
Norwood, MA 02062
Phone: 781-769-9750
Fax: 781-769-6334
e-mail: artech@artechhouse.com

Artech House
46 Gillingham Street
London SW1V 1AH UK
Phone: +44 (0)20 7596-8750
Fax: +44 (0)20 7630-0166
e-mail: artech-uk@artechhouse.com

Find us on the World Wide Web at: www.artechhouse.com